Norbert Müller
Perunthiruthy K. Madhu (eds.)

Current Developments in Solid State NMR Spectroscopy

With a Foreword by Richard R. Ernst

SpringerWienNewYork

Dr. Norbert Müller
Institute of Chemistry, Johannes Kepler Universität,
Linz, Austria

Dr. Perunthiruthy K. Madhu
Department of Chemistry, University of Southampton,
Southampton, UK

Cover illustration:
A. Goldbourt and P. K. Madhu, „Multiple Quantum ...“, Fig. 4, page 26, this volume
Typesetting: Thomson Press Ltd., Chennai, India
Printing: Manz Crossmedia, A-1051 Wien
Binding: Papyrus, A-1100 Wien
Printed on acid-free and chlorine-free bleached paper

SPIN: 10898512

With 74 Figures and Tables

CIP data applied for

Special Edition of
Monatshefte für Chemie/Chemical Monthly Vol. 133, No. 12, 2002

ISBN 3-211-83894-5 Springer-Verlag Wien New York

Editorial

NMR has now been established as an indispensable tool for structure elucidation and imaging. While the first two Nobel prizes for NMR scientists have been to the discovery of NMR and the development of various techniques that enabled wide-spread application of NMR experiments, the recent (2002) Nobel prize for chemistry to Kurt Wüthrich emphasizes the impact of NMR on biomolecular structure determination. Such a wide spread use of NMR experiments for structure elucidation, especially of proteins, has so far mainly been possible in isotropic solutions.

In contrast to the situation in liquid-state NMR, many theoretical and technological issues in solid-state NMR were clarified or resolved only in the last decade and by now libraries of well defined pulse sequences for solid-state NMR are accessible to a large number of scientists in different application areas. Although many of the theoretical foundations for this progress were established much earlier, the general understanding of the fundamentals of solid-state NMR, namely properties of internal spin Hamiltonians and how to manipulate them to obtain the desired spectroscopic information, have until recently lagged behind the situation in liquid-state NMR.

A large variety of systems can be studied using solid-state NMR. They range from zeolites, ceramics, and industrial polymers to biological systems, like membrane proteins. For atomic resolution structural investigations of amorphous and semi-crystalline materials as well as membrane proteins solid-state NMR techniques enjoy the status of a quasi-monopoly.

About 70% of the periodic table is made up of quadrupolar nuclei, which have previously been rather difficult to observe and manipulate in NMR. These elements are found or used in many biological systems and industrial materials. NMR studies of such nuclei have recently made a big leap forward with the inception of multiple-quantum magic-angle spinning (MQ-MAS). The large number of papers that has since then appeared underscore the importance of quadrupolar nuclei in general and the potential of the MQ-MAS approach.

Numerical simulation of spin physics under the action of pulsed radio frequency irradiation and rapid sample rotation is an indispensable prerequisite for the design of pulse sequences and for the interpretation of the spectral response in terms of structure and dynamics. Due to the complexity of nuclear spin interactions in the solid-state numerical simulations are required to find guidelines for designing tailored pulse sequences for specific problems. The wide spread demand for numerical simulations triggered development of universal NMR programming environments. Such numerical simulation packages act like virtual spectrometers enabling the simulation of experiments prior to their actual implementation on real instruments. This not only speeds up the design and optimization of new pulse sequences in solid-state NMR but also allows the design to be independent from instrumental imperfections, anticipating and actually promoting improvements in NMR hardware.

Solid-state NMR is a rapidly developing field of science. However the final directions emerging from these developments are not yet clearly defined, although some application areas (like heterogeneous catalysts and biological membranes) seem to benefit extraordinarily and have become driving forces of future development. The emerging applications of solid-state NMR in key research areas like proteomics and new materials have created a scenario which is reminiscent of the situation in liquid-state NMR around 1980, when pulse-FT NMR had become routine in chemistry laboratories and two-dimensional techniques were rapidly adopted by a large number of scientists. Today a variety of elegant techniques have been established in solid-state NMR and have already been implemented on state-of-the-art commercial NMR spectrometers.

We are currently witnessing a remarkable convergence of solid-state and liquid-state NMR theory and methodology in complementary ways. For example, interactions which are normally not observable in liquid-state NMR, like dipolar couplings and chemical shift anisotropies, are detected after partially aligning the solute molecules with the static magnetic field by suitable liquid crystalline solvent systems. Concurrently re-coupling pulse schemes achieve a similar goal in solid-state NMR by restoring interactions that are normally removed by magic-angle spinning, MAS. Many physiologically relevant questions, involving structures at phase interfaces will benefit significantly from a concerted approach by liquid-state and solid-state NMR methodology. While many newly developed solid-state NMR methods try to mimic the information display and apparent simplicity of well established liquid-state techniques, other pulse schemes exploit the unique properties of solid-state spin Hamiltonians to extract unique information pertinent to the solid phase. These developments are progressing at a rapid pace and the mutual inspiration of theory and applications will likely bear fruit even in unexpected areas of science and technology.

In this special edition of "Monatshefte für Chemie/Chemical Monthly" several of the issues mentioned above have been addressed. There are articles that extensively deal with issues of quadrupolar spins, decoupling techniques in spin-1/2 systems, simulation software, and applications. They represent a collection of snap shots of the state-of-the-art in important realms within the emerging key technology of solid-state NMR spectroscopy.

Norbert Müller
Perunthiruthy K. Madhu

Foreword

Why to apply solid-state NMR? – By now, we should have learned that NMR is mainly used for the study of molecules in solution, while x-ray diffraction is the method of choice for solids. Based on this fact, the two recent 'NMR-Nobelprizes' went indeed into the liquid phase: my own one eleven years ago, and particularly the most recent one to *Kurt Wüthrich*. His prize is beyond any doubts very well justified. His contribution towards the study of biomolecules in solution, in their native (or almost native) environment is truly monumental. We all will profit from it indirectly when one of our future diseases will be cured with better drugs, based on the insightful knowledge gained through liquid-state NMR.

Two fields of NMR are still left out of the Nobel Prize game: magnetic resonance imaging (MRI) and solid-state NMR. The disrespect for MRI in Stockholm is particularly difficult to understand; but this is not a subject to be discussed at the present place. Solid-state NMR is the third of the three great fields of NMR, powerful already today and very promising for the near future.

Already *Cornelis Gorter* experienced in 1937, painfully, how difficult NMR in the solid phase can be. Indeed, he failed even to detect a signal. The developments in solid-state NMR were slower than in liquid phase, not because of the lesser interest but because of the much larger hurdles to be overcome. It is at very first the apparently restricted resolution of NMR in the solid, due to the infinite network of dipolar interactions and due to the anisotropies, and the limited sensitivity that might frighten potential users. Today, most of these inherent difficulties have been mastered and the tool is ready to be applied very fruitfully, perhaps again to impress sooner or later should the Committee in Stockholm.

Still, the question remains: Why should NMR in the solid-state be applied instead of the extremely powerful x-ray crystallography? – There are indeed good reasons: First, many interesting molecules can not be properly crystallized to permit an investigation by x-ray diffraction, and, more importantly, most natural materials are not occurring in ideal crystalline form. Whenever there is disorder, NMR might be the method of choice. The second reason is that most materials are inherently of a dynamical nature. Intra-molecular and intermolecular dynamics is not only a fact but largely determines the beneficial properties of technological and biological materials. NMR can handle this situation masterfully and can deliver much essential information on dynamics in the solid state. Thirdly, not all solid-like systems are truly 'solid'. There is a vast field of mesophases, somewhere between a liquid and a 'dead' solid. Liquid crystals, gels, micelles, and tissues belong to the fast class of mesophases. They again form a beautiful 'playground' for NMR, which is perhaps the best conceivable method for the study of these partially disordered systems. Indeed most of the solid-like systems of relevance in biology and biomedicine have attributes of inherent disorder: tissues, membranes, bones, you name it. And solid-state NMR has still an enormous potential for future applications in the life sciences.

Many of the advanced technological organic and inorganic materials are disordered and dynamical, as well. Just think about synthetic polymers. Indeed, much of the detailed knowledge we have today on polymeric materials stems from advanced NMR investigations. Interesting systems, combining life sciences with polymeric properties, are the technologically important natural fibers, such as silk and the various fibers of plant origin. There is also a vast class of inorganic, partially disordered, systems suitable for NMR studies, such as zeolites that of great technical relevance. Here, molecular adsorption phenomena are of relevance where again dynamical features are crucial.

Truly, there is a vast field of highly relevant applications of solid-state NMR for the future. I am sure that additional mental and financial investments into this exciting field are well justified and are needed to solve the great problems posed by our all destiny.

Richard R. Ernst
Nobel Laureate in Chemistry 1991
Laboratorium für Physikalische Chemie
ETH Hönggerberg HCI, Zürich
October 2002

Contents

Invited Reviews

Contributions

Invited Review

Dipolar and Scalar Couplings in Solid State NMR of Quadrupolar Nuclei

Alexej Jerschow*

Department of Chemistry, New York University, New York, NY 10003

Received April 16, 2002; accepted May 15, 2002
Published online October 7, 2002

Summary. Most NMR-active nuclei found in the periodic table have a quadrupole moment. In combination with a nonsymmetric electron distribution a strong NMR-active interaction results, which very often overshadows the dipolar and scalar couplings. This article aims at reviewing how these interactions manifest themselves in quadrupolar NMR and how they can be exploited for resonance assignment and structure elucidation, in spite of the presence of a strong quadrupolar interaction.

Keywords. Solid-state; NMR spectroscopy; Dipolar couplings; Quadrupolar couplings.

Introduction

Most NMR-active nuclei in the periodic table have a quadrupole moment, that is a nonsymmetric nuclear charge distribution. In combination with a nonsymmetric electron distribution a strong NMR-active interaction results, which very often overshadows other interactions such as dipolar and scalar couplings. It is on the basis of observing and interpreting manifestations of these interactions that structure elucidation has been extremely successful in spin $1/2$ NMR in one and multiple dimensions in both the liquid and the solid states [1, 2]. Dipolar couplings allow the determination of interatomic distances *via* NOESY-type experiments [1, 3] in the liquid state and dipolar recoupling techniques [4–6] and dipolar correlation experiments [7] in the solid state. Correlation spectroscopy *via* scalar J couplings is routinely applied in the liquid state (COSY-type experiments) and has likewise been demonstrated in solid state NMR [8–11] (see Ref. [12] for a recent comprehensive review).

In solid-state NMR of quadrupolar nuclei these approaches have not been as successful because the dipolar and scalar couplings are often negligible compared

* E-mail: alexej.jerschow@nyu.edu

to the dominant quadrupolar coupling. Notable exceptions are cases where the quadrupolar couplings are comparatively weak, such as *e.g.* in ^{6}Li and ^{7}Li NMR.

This article aims at reviewing how dipolar and scalar couplings manifest themselves in solid state quadrupolar NMR and how they can be exploited for resonance assignment and structure elucidation in cases where the quadrupolar coupling is large compared to these interactions.

One-dimensional NMR

Because of the strength of the quadrupolar coupling interaction, which may be on the order of several MHz, dipolar and scalar couplings have rarely been observed in static one-dimensional NMR experiments. The strength of the dipolar coupling interaction may be on the order of hundreds of Hz up to a kHz (*e.g.* 532 Hz for Mn–Mn with a distance of 2.4 Å), which is minute compared to the quadrupolar interaction (*e.g.* a linewidth of 2 MHz due to a typical quadrupolar coupling strength covers 100 times the full ^{13}C chemical shift range at a 100 MHz *Larmor* frequency). Heteronuclear dipolar coupling constants to protons may be as large as 10 kHz (*e.g.* 9.3 kHz for ^{27}Al–^{1}H with a 1.5 Å distance), but are nonetheless significantly weaker than common quadrupolar couplings. Reports of observations of these couplings can therefore be expected to be scarce. For half-integer spins, where the central transition is free from the first order quadrupolar broadening [13] it should, however, be straightforward to observe dipolar as well as scalar couplings in an additional linebroadening and linesplitting of the central transition lines, since the second order quadrupolar broadening is on the order of just a few kHz. Figure 1 shows experimental and simulated static powder lineshapes for [^{17}O] and [^{13}C, ^{17}O]benzamide. The former displays a usual second order powder lineshape expected for the central transition of a quadrupolar nucleus. The latter shows additional splitting resulting from the dipolar couplings between the ^{13}C and the ^{17}O nuclei. In this case these additional lineshape effects were used to determine the absolute orientations of the electric field gradient tensor components and the internuclear ^{13}C–^{17}O vector.

First Order Effects

Magic angle spinning (MAS) has been employed successfully in NMR of spin $1/2$ nuclei to obtain high resolution spectra. In quadrupolar NMR this method also leads to significant line narrowing due to (a) the removal of the first order quadrupolar coupling and (b) the reduction of the second order quadrupolar coupling (see below). One would think that here the dipolar couplings should be easy to observe, since one only needs to compete with the second order quadrupolar broadening (on the order of a few kHz). The dipolar couplings are, however, likewise averaged away rather efficiently by MAS. The strength and orientation parameters of the dipolar couplings can in theory (in theory only) be determined by analysing the intensities of the spinning sidebands [14]. The quadrupolar couplings also contribute to the intensities and due to the relative weakness of the dipolar couplings this approach seems prohibitive. It should be mentioned, however, that analysing the spinning sidebands in the spectra of spin $1/2$ nuclei coupled to quadrupolar nuclei is

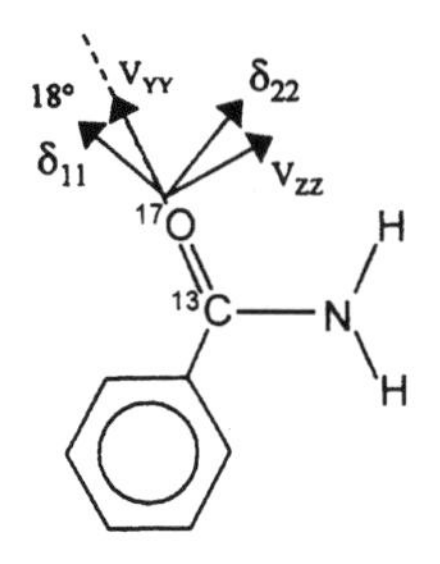

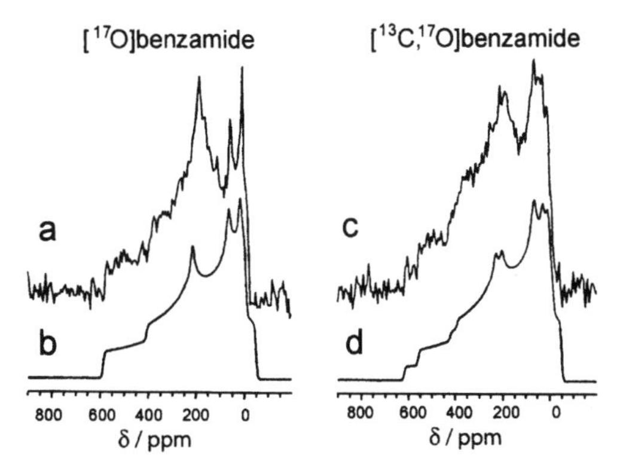

Fig. 1. Experimental and simulated static powder lineshapes for [^{17}O]benzamide (a, b) and [^{13}C, ^{17}O]benzamide (c, d). The additional linesplitting in (c) and (d) arises from the dipolar couplings between the ^{13}C and the ^{17}O nuclei (reproduced from Ref. [81], with permission)

a straightforward procedure that has provided useful structural information based on dipolar couplings [15–17].

Scalar coupling (the isotropic part), on the other hand should still be observable as it is independent of orientation. Its small strength (up to a few hundred Hz), however, makes its observation rather difficult. Figure 2 shows one of the few documented observations of scalar coupling in a spectrum of a quadrupolar nucleus [18]. While the coupling has a perceptible effect on the powder lineshape it should be noted that in this particular case the presence of the isotropic J coupling has been established by other means (see below). Again, scalar couplings between 1/2 and quarupolar nuclei are best observed in the spectrum of the spin 1/2 nucleus [19].

Second Order Effects

It is well known that MAS NMR spectra of spin 1/2 nuclei coupled to quadrupolar nuclei (*e.g.* ^{14}N) show peculiar non-symmetric lineshapes and linesplittings, which can be explained by a second order dipolar coupling (a cross-term between the quadrupolar and the dipolar coupling) [20, 21]. This additional linebroadening is not averaged away by MAS. It scales as $\omega_Q\omega_D/\omega_0$, where $\omega_Q = C_Q/2I(2I-1)$, with C_Q the nuclear quadrupolar coupling constant, ω_D the dipolar coupling constant,

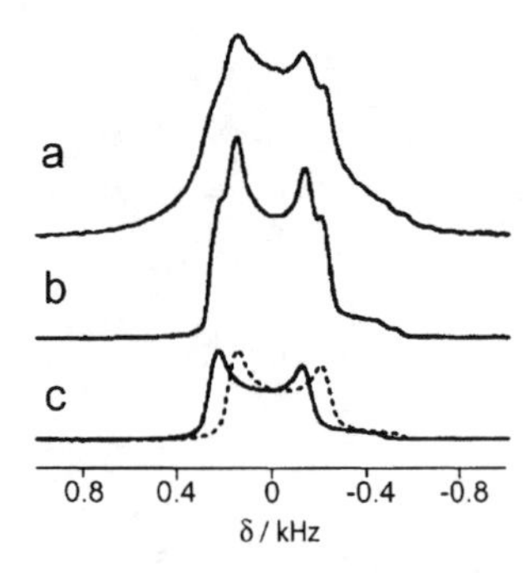

Fig. 2. ^{11}B MAS spectra of $(PhO)_3P{-}BH_3$ obtained at 7.4 T. (a) Experimental spectrum, (b) simulated spectrum, and (c) simulated subspectra for the two components of the doublet arising from J coupling to ^{31}P. A coupling constant of approximately 85 Hz was extracted (reproduced from Ref. [18], with permission)

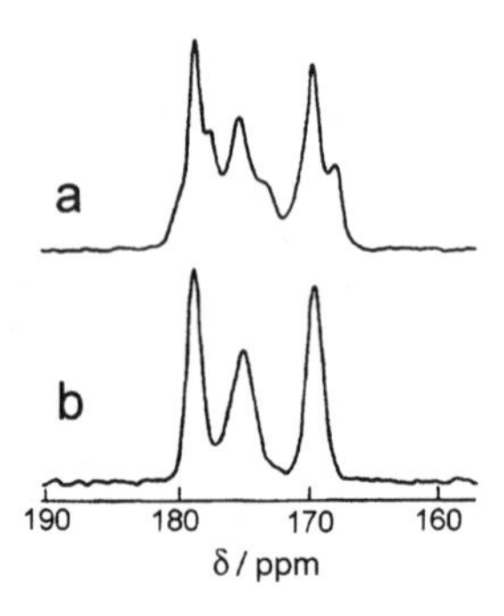

Fig. 3. ^{13}C MAS spectra showing (a) residual second order dipolar lineshape distortions arising from coupling to ^{14}N, and (b) their removal by isotopically enriching the molecule with ^{15}N. The spectra arise from the cyanide carbons of the coordination polymer $[(Me_3Sn)_4Fe(CN)_6]_\infty$. There are three cyanide environments in the crystal structure (reproduced from Ref. [21], with permission)

and ω_0 the *Larmor* frequency, given in angular velocity units. The resonance broadening and splitting effects can be on the order of up to a few hundred Hz. Figure 3 shows a ^{13}C MAS spectrum where the residual second order dipolar coupling to ^{14}N leads to pronounced lineshape distortions. The authenticity of the lineshape effects is tested by acquiring the same spectrum from an ^{15}N labeled sample, where the distortions disappear. Similar effects have been observed for many nuclear pairs, including ^{31}P–$^{63/65}Cu$, ^{31}P–^{55}Mn, ^{119}Sn–$^{35/37}Cl$, ^{13}C–^{23}Na, ^{29}Si–^{14}N, and even ^{13}C–^{2}H [21]. In the latter case the quadrupolar couplings are comparatively weak, but the larger dipolar couplings associated with the deuterium nucleus compensate for this [22].

In a MAS NMR spectrum of a quadrupolar nucleus, where the linewidth resulting from the second order quadrupolar coupling can be a few kHz in size, these effects are also expected to be visible. Likewise, homonuclear coupling between two quadrupolar nuclei should have pronounced effects. Because of the complicated orientational dependencies involved (the second order dipolar/quadrupolar

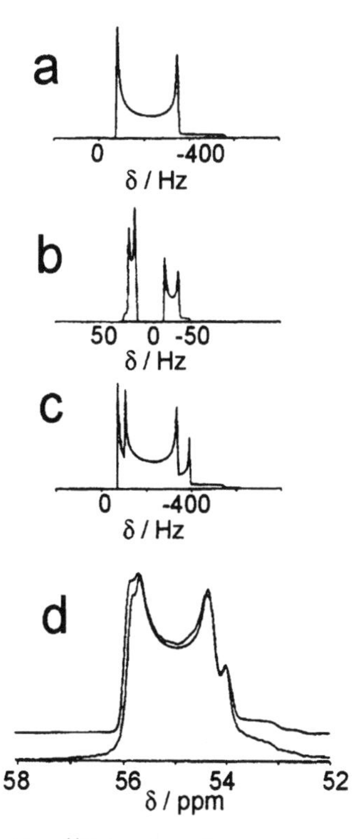

Fig. 4. Simulated central transition ^{11}B MAS NMR spectra arising from the second-order ^{11}B quadrupolar interaction (a), the residual second order $^{11}B-^{14}N$ dipolar interaction alone (b), and both interactions (c). (d) Experimental ^{11}B MAS NMR spectrum of triethanolamine borate obtained at 11.75 T (adapted from Ref. [24], with permission)

cross-term depends on the strength and orientations of the quadrupolar tensor, as well as, on the orientation of the internuclear vector and the distance between the nuclei) additional lineshape effects, though sufficiently strong, may not be immediately apparent in a one-dimensional MAS spectrum of a powder sample (see for example Ref. [23]). Figure 4 shows simulations and experimental evidence of lineshape distortions of the central transition signal of ^{11}B [24] in the presence of a residual dipolar coupling to ^{14}N.

Dipolar Recoupling

The (first order) dipolar interaction is normally averaged out over a rotor cycle by MAS (see Fig. 5). However, if this time averaging is interrupted by the application of rotor-synchronized rf irradiation, a non-vanishing effective dipolar coupling arises (Fig. 5b) [4, 5]. Heteronuclear dipolar interactions have been recoupled in this way for spin $1/2$ nuclei by a technique called REDOR (Rotational Echo Double Resonance), which uses two π pulses per rotor cycle [6]. Approximately 70% of the dipolar coupling can be recoupled using REDOR. This coupling leads

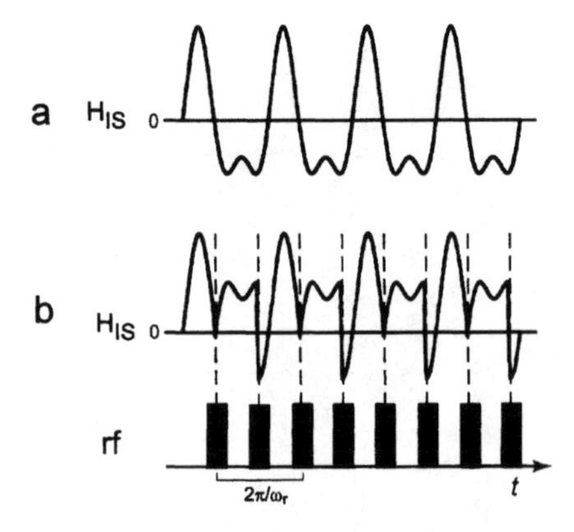

Fig. 5. The dipolar interaction is averaged away by MAS. (a) Shows an example of the change of the magnitude of the heteronuclear dipolar Hamiltonian during the rotation for a particular crystallite orientation. (b) Shows how the averaging process can be interrupted by the application of 180 degree pulses during the rotor period. An average dipolar interaction results

to a signal dephasing, which is recorded after different recoupling times. A "dephasing curve" is obtained which can be used to extact the dipolar coupling constants [6]. The pulse sequence of the REDOR experiment is shown in Fig. 6a.

This approach has also been successfully implemented for determining distances between quadrupolar and spin 1/2 nuclei, and even between different quadrupolar nuclei (*e.g.* ^{11}B and ^{27}Al) [25, 26]. In many cases, however, REDOR is not very efficient since it relies on the possibility of complete magnetization inversion for a quadrupolar nucleus. This is rarely possible. Another approach, called TRAPDOR (transfer of populations in double resonance), has been shown to give better sensitivity to dipolar couplings. In this case a continuous rf field is applied to the quadrupolar nucleus and several level transitions occur adiabatically over a rotor cycle (Fig. 6b shows the pulse sequence). While REDOR affects for the most part only the $\pm\frac{1}{2}$ states, TRAPDOR affects all levels and therefore leads to more effective dephasing [27–29]. In this way decay curves can be acquired, which are characteristic of the dipolar and quadrupolar couplings. While the extraction of quantitative information relies on computer simulations and is sometimes cumbersome, this technique has been very useful to detect nuclei with large quadrupolar couplings "indirectly". This is done by monitoring the decay of the magnetization of a spin 1/2 nucleus coupled to the quadrupolar spin as a function of the offset of the rf field applied to the quadrupolar spin [30, 31]. Two spectra need to be acquired, one with rf irradiation at the *Larmor* frequency of the quadrupolar nucleus, and one without. The difference spectrum shows the spectrum of the nuclei coupled to the quadrupolar nucleus. An example is shown in Fig. 7, where the signals from the ^{31}P nuclei neighboring ^{27}Al nuclei were filtered out.

Sometimes the TRAPDOR experiment can become difficult to perform because of rf power and irradiation time limits. The REAPDOR experiment (Rotational echo, adiabatic passage, double resonance) can prove more useful in these cases. Instead of a continuous wave irradiation on the quadrupolar nucleus for a duration of several rotor cycles (up to several milliseconds) a spin-lock pulse is applied for

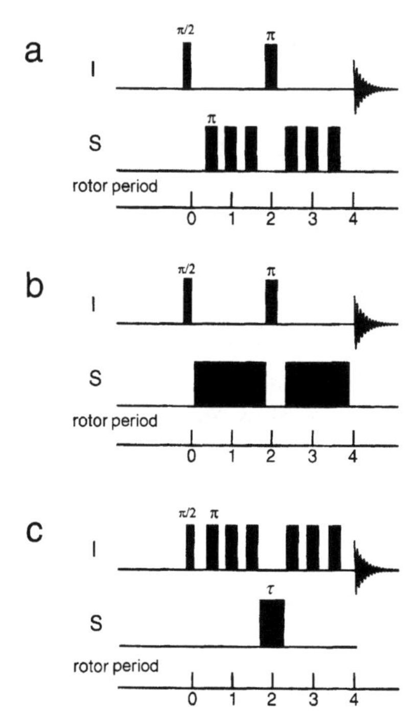

Fig. 6. (a) The REDOR pulse sequence. Two 180 degree pulses per rotor period recouple the dipolar coupling. (b) The TRAPDOR sequence. A continuous wave field applied to the quadrupolar nucleus (S) over several rotor periods recouples the dipolar coupling. (c) The REAPDOR sequence. The dipolar coupling is recoupled by two 180 degree pulses per rotor period on the spin 1/2 nucleus (I) and a continuous wave field on the quadrupolar nucleus (S) for approximately 1/3 of a rotor period

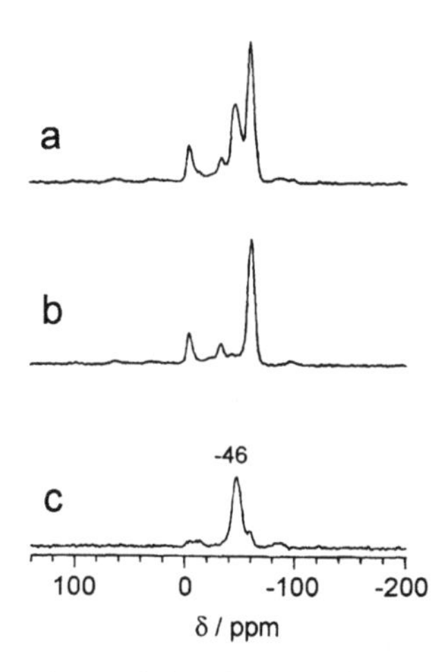

Fig. 7. $^{31}P/^{27}Al$ TRAPDOR experiment of trimethylphosphine on dehydroxylated zeolite HY. Without ^{27}Al irradiation (a), with irradiation (b), and the TRAPDOR spectrum (c) which is the difference between (a) and (b) (reproduced from Ref. [30], with permission)

only a fraction of a rotor period (Fig. 6c shows the pulse sequence). The rest of the time 180 degree pulses on the neighboring spin $1/2$ nucleus prevent the dipolar coupling from refocusing. At the same time the effects of the chemical shift anisotropies are eliminated during the dephasing time. The spin-lock pulse on the quadrupolar nucleus partly inverts the populations of the quadrupolar nucleus. The optimal duration of the pulse needs to be adjusted individually but is approximately $1/3$ of the rotor period. Without it the second half of the pulse sequence would exactly retrace the evolution of the first half, and no dipolar dephasing would occur. This represents a reference experiment, accounting for T_2 relaxaiton. REAPDOR has been used, for example, to determine ^{13}C–^{17}O and ^{13}C–^{14}N distances [28, 32, 33].

For homonuclear dipolar couplings between two quadrupolar nuclei different techniques have been demonstrated, which will be discussed in the next section.

Correlation Spectroscopy

In NMR of spin $1/2$ nuclei correlation spectroscopy has been extremely useful in the assignment and interpretation of spectra. Spatial proximity can easily be probed, as well as bonding through hetero- and homonuclear scalar couplings in both the liquid and the solid states [1–3, 7–9, 34]. For quadrupolar nuclei similar solid-state NMR approaches have been close to inexistent until recently. This section reviews some of the recent developments in this field.

High-Resolution Techniques

Techniques for achieving high-resolution spectroscopy of quadrupolar nuclei have been known for a long time. This was originally achieved by rotating the sample simultaneously around two axes (double rotation – DOR [35]), which removes both the first and the second order quadrupolar coupling. Another approach, which is less demanding on hardware is to rotate the sample around different axes [36] (dynamic angle spinning – DAS). As a result one obtains anisotropic–isotropic correlation spectra. The isotropic dimension shows resonances free from quadrupolar broadening effects. While these experiments have enormous potential, the practical implementations are often cumbersome due to the mechanical limitations of rapidly spinning around two different axes. In 1995 a method was developed, called multiple quantum magic angle spinning (MQMAS), which allows one to use standard hardware to obtain anisotropic–isotropic correlation spectra [37, 38]. These two-dimensional experiments show an isotropic dimension with resonances free from the first and second order quadrupolar broadenings, correlated with second order anisotropic lines in a second dimension. J couplings can therefore be resolved in this way. In particular the $J = 85 \pm 5$ Hz coupling constant between ^{11}B and ^{31}P in $(PhO)_3P$–BH_3 has been determined and was used to unravel the components from the J doublet in the powder spectrum of Fig. 1 [24].

It has since been shown that the narrow resonances in the isoptropic dimension show another fine structure that is related to hetero- and homonuclear through-space (dipolar) couplings [24, 39–42]. This shows up in both a linebroadening and a linesplitting. In the case of the dipolar coupling (and the anisotropic J coupling)

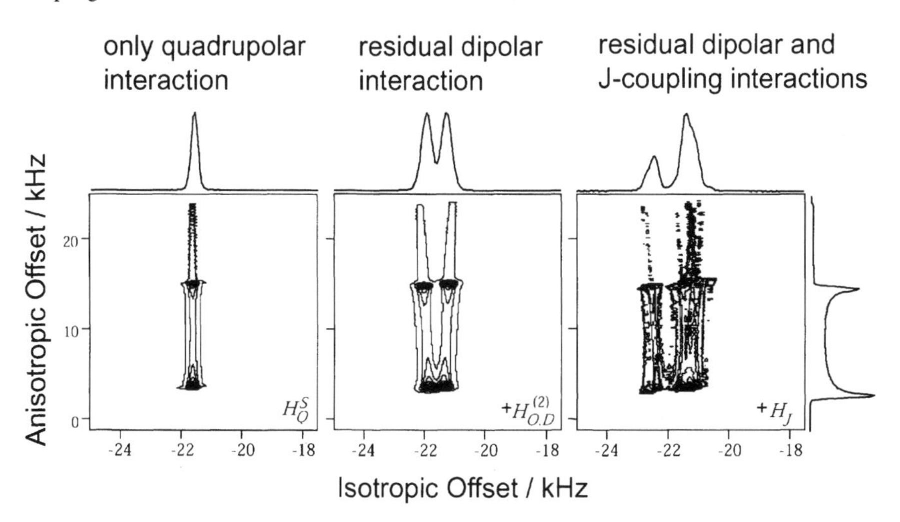

Fig. 8. Simulated 2D MQMAS NMR spectra with the quadrupolar interaction alone (a), the addition of the second order dipolar coupling (b), and the scalar coupling (c). The spectra were calculated for a pair of equivalent ^{11}B nuclear spins ($I = 3/2$) with 1.5 Å distance in a field of 4.7 T with a quadrupolar coupling constant $C_Q = 5$ MHz, a J coupling constant of 100 Hz, and a natural line width of 20 Hz (reproduced from Ref. [40], with permission). It should be noted that in this representation the isotropic dimension (narrow resonances) is along the horizontal axis, while the anisotropic dimension (broad resonances) is along the vertical axis. Many authors use the opposite representation

this can be explained by the same second order effect as discussed in Section 2.2. The isotropic J coupling produces a first order effect since it is not removed by MAS. For this reason, the second order dipolar coupling ($\omega_D\omega_Q/\omega_0$) is of the same order of magnitude as the J coupling (up to a few hundred Hz). Figure 8 shows an example of the fine structure observed in MQMAS spectra with both dipolar and scalar couplings present. One can therefore extract dipolar and scalar coupling parameters *via* iterative fitting to the experimental data. This process should be more accurate than fitting one-dimensional spectra, because the resonance broadening effects are not hidden by the large quadrupolar coupling.

Recently, a new type of MQMAS experiment has been presented, which produces high-resolution NMR spectra of quadrupolar nuclei by combining two different multiple quantum coherences [43]. By using different combinations of multiple quantum coherences in the experiments, one obtains different chemical shift (and quadrupolar isotropic shift) scaling factors. While heteronuclear interactions scale in exactly the same way as the chemical shift, homonuclear interactions scale differently. It is therefore expected, that combining different multiple quantum coherences for a given spin can help to increase the accuracy of iterative fits.

MQMAS spectra also display spinning sidebands in the isotropic dimension. Extraction of dipolar coupling parameters, however, appears to be difficult, since the major contribution to the sideband intensities comes from the first order quadrupolar coupling in combination with the creation and conversion efficiencies of multiple quantum coherences [44].

MQMAS has been combined with REDOR, where the dipolar dephasing sequence was applied during the multiple-quantum evolution [45]. The authors found that the sensitivity of this method for the determination of dipolar couplings is higher than for a sequence where the dipolar dephasing occurs during single quantum evolution. This is due to the fact that the heteronuclear dipolar coupling affects p quantum coherence p times stronger than single quantum coherence, hence the dephasing effect is stronger. This experiment may be a good alternative to REAPDOR and TRAPDOR in cases where efficient inversion of the multiple quantum coherence is possible.

Other, potentially more advanced methods for generating high-resolution spectra for quadrupolar nuclei have been proposed, involving the correlation of satellite transitions with central transitions [46–48]. While the the peaks are broadened significantly in the isotropic dimension by inaccuracies in the setting of the magic angle, a new type of split-t_1 experiment proposed by *Wimperis* and coworkers may represent the ultimate solution to this problem [49]. It is expected that dipolar and scalar coupling interactions will be well perceptible in these experiments.

Cross-Polarization, Heteronuclear Correlation

In spin 1/2 NMR, a technique called cross-polarization (CP) is widely used to transfer magnetization from one nuclear species (usually the abundant nuclei, such as ^1H) to another (usually the rare spin, such as ^{13}C) [50–52] *via* the dipolar or the scalar couplings between those nuclei. The magnetization transfer is facilitated by the fact that the nutation frequencies of both spins are made equal by properly adjusting the respective radiofrequency fields such that $\omega_{1I} \approx \omega_{1S}$ (this is the so-called *Hartmann-Hahn* match condition [53]).

In the presence of the quadrupolar interactions the above considerations cannot be used directly to perform an efficient polarization transfer, since for many crystallites in a powder sample one has $\omega_Q \gg \omega_1$ (ω_1 is the radiofrequency field strength). The matching condition can be calculated directly in this limit [54, 55] as

$$\omega_{1I} = (S + 1/2)\omega_{1S}. \tag{1}$$

Under this condition the polarization is transferred from the I nucleus to the S (the quadrupolar) nucleus [54]. This technique can be combined with MQMAS [56, 57]. One can also transfer the polarization directly to multiple quantum coherences by using slightly modified matching conditions [55]. These experiments have also been combined with MQMAS and up to seven quantum coherence has been obtained using CP [58–62]. In the general case the sample contains also crystallites with $\omega_1 \approx \omega_Q$, so that there is no longer a clear distinction between the different matching conditions, and a suitable phase cycle needs to be used to select the proper coherence order [63].

Contrary to the situation in spin 1/2 NMR, CP in quadrupolar NMR does not in general offer signal enhancement but is more appropriately used for correlation spectroscopy and for spectral editing techniques, *i.e.* filtering out signals from nuclei that are close to spin 1/2 nuclei. Figure 9 shows an example of a correlation spectrum that can be obtained using a combination of CP and MQMAS.

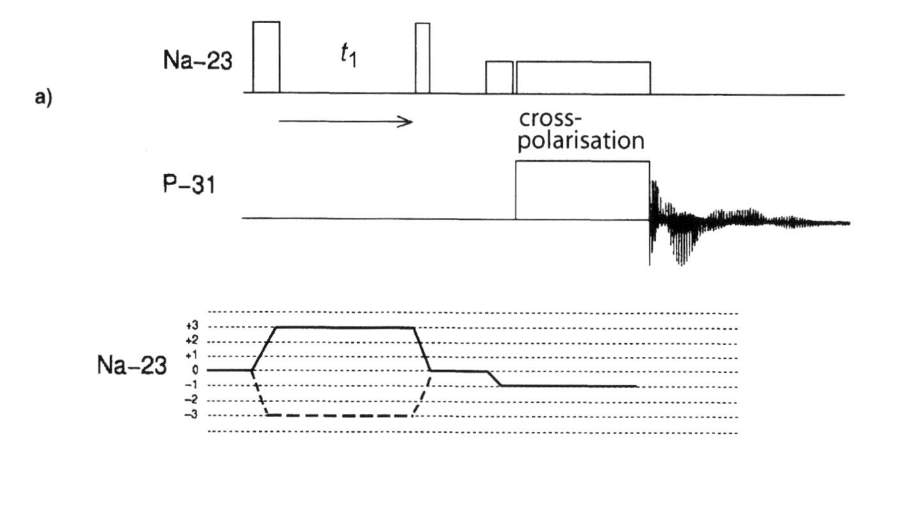

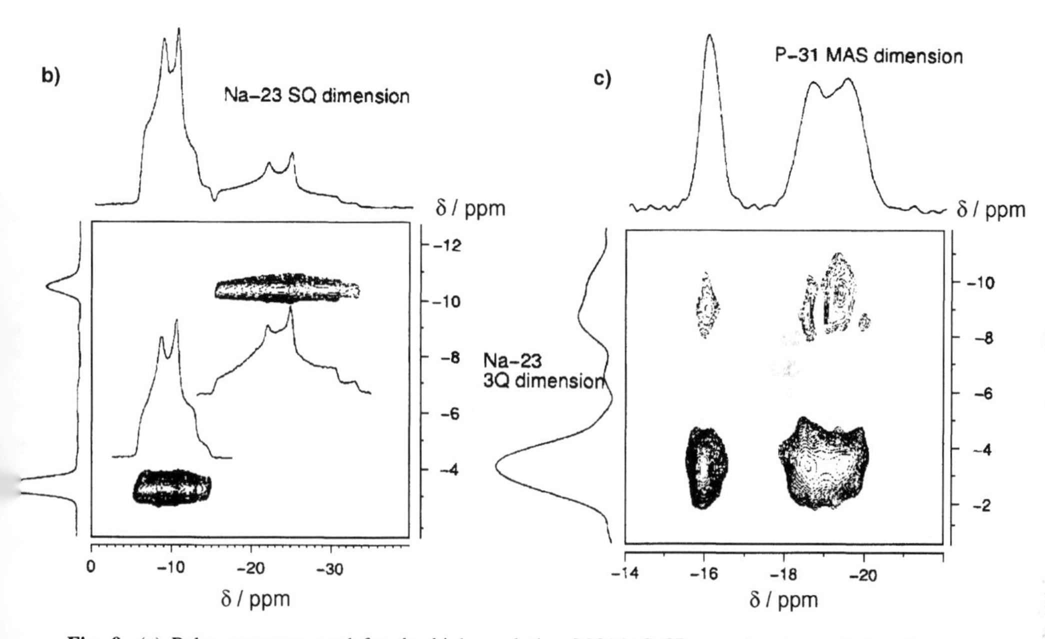

Fig. 9. (a) Pulse sequence used for the high-resolution MQMAS-CP experiment correlating the isotropic frequencies of the quadrupolar nucleus (S) with the frequencies of a spin $1/2$ nucleus *via* CP. (b) The ^{23}Na 3QMAS spectrum correlating isotropic and anisotropic ^{23}Na frequencies and (c) the result of a ^{31}P–^{23}Na correlation experiment on the compound $Na_3P_3O_9$ (reproduced from Ref. [66], with permission)

Some signal enhancement may be obtained by using "reverse" CP to transfer the polarization from the quadrupolar nucleus to spin $1/2$ nucleus [64–66]. In combination with MQMAS one can obtain two-dimensional correlation experiments with

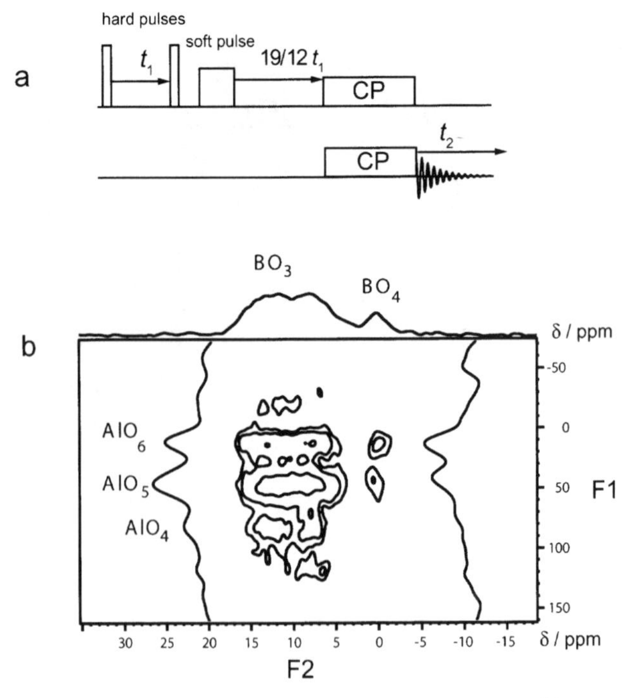

Fig. 10. (b) 2D correlation experiment between the isotropic frequencies of ^{27}Al and the anisotropic frequencies of ^{11}B. The pulse sequence for this experiment is shown in (a) (similar to the one in Fig. 9a; reproduced from Ref. [69], with permission)

an isotropic frequency dimension for the quadrupolar nucleus and a chemical shift dimension for the spin $1/2$ nucleus. Figure 9c shows a high-resolution ^{23}Na–^{31}P correlation spectrum obtained from a sample of $Na_3P_3O_9$ [66].

The matching conditions for CP between two different quadrupolar nuclei (in this case half-integer nuclei) are yet different [67–69], and a heteronuclear correlation experiment between ^{11}B and ^{27}Al has been presented (shown in Fig. 10).

Homonuclear Recoupling

In addition to the heteronuclear dipolar interactions, the homonuclear dipolar interactions can be recoupled for the purpose of measuring distances and performing correlation experiments. By applying a properly tuned continuous wave radiofrequency field or a multiple pulse sequence it is possible to interrupt the averaging of the dipolar interactions that happens due to MAS and to recouple parts of the dipolar coupling Hamiltonian. In the quadrupolar case this technique has to be used with caution, since the same pulse sequence that recouples the homonuclear coupling will also recouple the quadrupolar coupling. Since the latter is much stronger it would lead to a quick decay of the magnetization leaving little hope that the recoupled dipolar coupling would show a perceptible effect. *Griffin* and coworkers [70] have shown that if a continuous wave field is applied which is

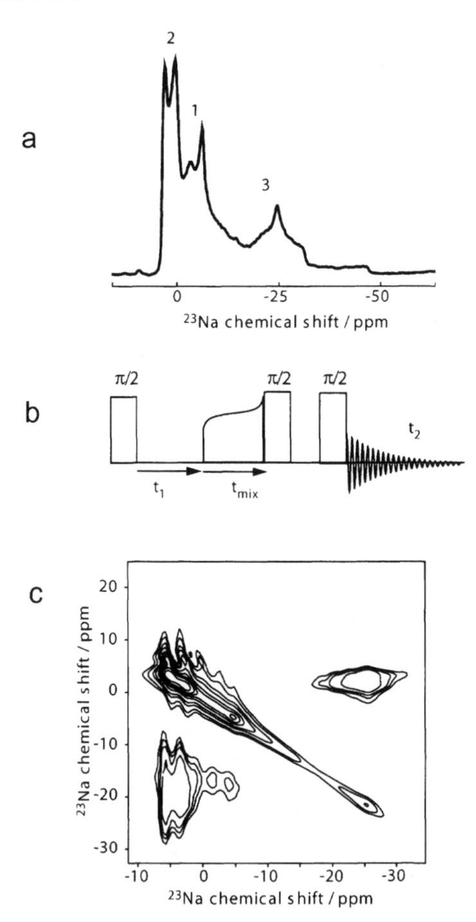

Fig. 11. (a) A one-dimensional MAS ^{23}Na spectrum indicating the three crystallographically different sites. (b) Pulse sequence and (c) spectrum for a homonuclear correlation experiment between the three crystallographically distinct ^{23}Na sites in Na_2HPO_4 (reproduced from Ref. [70], with permission)

sufficiently weak compared to the spinning speed one can keep the magnetization of the central transition locked (*i.e.* the effect of the quadrupolar coupling is minimized), while the dipolar coupling is recoupled sufficiently so that a transfer of the magnetization occurs. This has been used for a homonuclear correlation experiment between the three crystallographically distinct ^{23}Na nuclei in Na_2HPO_4, which is shown in Fig. 11.

Meier and coworkers further showed that the magnetization can be transferred using a technique commonly known as rotational-resonance [71], and a correlation spectrum has been obtained in this way [72]. The rotational-resonance condition is met when the difference between the resonance frequencies of the two nuclei matches an integer multiple of the spinning speed. In this case the magnetization may be exchanged between the two spins. Furthermore, the authors showed that an increase in polarization transfer can be obtained when the rotor speed is swept through the rotational-resonance condition, as in this case the matching condition is broader. This transfer has also been shown to be possible for deuterium [73].

For quadrupolar spin systems it has been shown that many more rotational resonance matching conditions can be found if a weak radiofrequency field is applied [74–76]. These effects can be exploited for enhancing the excitation efficiency in MQMAS experiments on the one hand, but more importantly can be used to create three-dimensional high-resolution correlation experiments based on homonuclear dipolar couplings [77].

For completeness, we also mention that by offsetting the sample spinning axis from the magic angle, one also reintroduces the dipolar couplings, which has been used to obtain a correlation spectrum for a ^{23}Na and ^{11}B-containing sample [78].

Other Methods

As a final note we would like to mention a curious correlation experiment that was recently presented by *Duer* and *Painter* [79]. In this experiment dipolar coupling between two $3/2$ quadrupolar nuclei (^{23}Na) is probed by creating six quantum coherence between them. This high quantum coherence can only be created when two of these spins are in close proximity. Formally, this experiment corresponds to the INADEQUATE experiment for spin $1/2$ nuclei, where double quantum coherence is created between the two nuclei [80].

Conclusions

In this article an overview was presented over manifestations of dipolar and scalar couplings in quadrupolar NMR. We mentioned linebroadening and linesplitting effects in one-dimensional NMR, methods for recoupling of the dipolar interaction in MAS spectra, residual couplings in high-resolution MQMAS experiments, and some of the more recent hetero- and homonuclear correlation experiments. The development of high-resolution methods has unveiled the underlying structure of the NMR interactions, which are normally completely hidden by the quadrupolar interactions. The focus on dipolar and scalar coupling interactions is a natural development in this area, as it is through these couplings that some of the most sophisticated NMR experiments provide structural information both in liquid and in solid state NMR today. It is probably only a matter of time until powerful hetero- and homonuclear solid-state correlation experiments of quadrupolar nuclei will become routine techniques.

References

[1] Cavanagh J, Palmer AG, Fairbrother W, Skelton N (1996) Protein Nmr Spectroscopy: Principles and Practice. Academic Press, San Diego
[2] Davis J, Auger M (1999) Prog Nucl Magn Reson Spectroscopy **35**: 1
[3] Wüthrich K (1996) The Encyclopedia of NMR, vol 2, Wiley, NY, pp 932–939
[4] Dusold S, Sebald A (2000) Annu Rep NMR Spectrosc **41**: 185
[5] Bennett AE, Griffin RG, Vega S (1994) NMR Bas Princ Progr **33**: 3

[6] Gullion T, Schaefer J (1989) J Magn Reson **81**: 196
[7] Rienstra CM, Hohwy M, Hong M, Griffin RG (1998) J Am Chem Soc **120**: 10602
[8] Fyfe CA, Grondey H, Feng Y, Kokotailo GT (1990) Chem Phys Lett **173**: 211
[9] Lesage A, Charmont P, Steuernagel S, Emsley L (2000) J Am Chem Soc **122**: 9739
[10] Hardy EH, Verel R, Meier BH (2001) J Magn Reson **148**: 459
[11] Heindrichs ASD, Geen H, Giordani C, Titman JJ (2001) Chem Phys Lett **335**: 89
[12] Laws DD, Bitter HML, Jerschow A (2002) Angew Chem Int Ed Engl (in press)
[13] Vega AJ (1996) The Encyclopedia of NMR, vol 4, Wiley, NY, pp 3869–3888
[14] Herzfeld J, Berger AE (1980) J Chem Phys **73**: 6021
[15] Davies NA, Harris RK, Olivieri AC (1996) Mol Phys **87**: 669
[16] Olivieri AC (1997) Solid State Nucl Magn Reson **10**: 19
[17] Ding S, McDowell A (1997) J Chem Phys **107**: 7762
[18] Wu G, Kroeker S, Wasylishen RE, Griffin RG (1997) J Magn Reson **124**: 237
[19] Asaro F, Camus A, Gobetto R, Olivieri AC, Pellizer G (1997) Solid State Nucl Magn Reson **8**: 81
[20] Hexem JG, Frey MH, Opella SJ (1982) J Chem Phys **77**: 3847
[21] Harris RK, Olivieri AC (1992) Prog NMR Spectrosc **24**: 435
[22] Jonsen P, Olivieri AC, Tanner SF (1996) Solid State Nucl Magn Reson **7**: 121
[23] Wi S, Frydman L (2000) J Chem Phys **112**: 3248
[24] Wu G, Yamada K (1999) Chem Phys Lett **313**: 519
[25] Chan JCC, Bertmer M, Eckert H (1999) J Am Chem Soc **121**: 5238
[26] Bertmer M, Züchner L, Chan JCC, Eckert H (2000) J Phys Chem B **104**: 6541
[27] van Eck ERH, Kentgens APM, Kraus H, Prins R (1995) J Phys Chem **99**: 16080
[28] Smith ME, van Eck ERH (1999) Progr Nucl Magn Reson Spectrosc **34**: 159
[29] van Wüllen L, Kalwei M (1999) J Magn Reson **139**: 250
[30] Kao H, Grey CP (1996) Chem Phys Lett **259**: 459
[31] Deng F, Yue Y, Ye C (1998) Solid State Nucl Magn Reson **10**: 151
[32] Gullion T (1995) Chem Phys Lett **246**: 325
[33] Gullion T (1995) J Magn Reson Series A **117**: 326
[34] Pauli J, van Rossum B, Förster H, de Groot HJM, Oschkinat H (2000) J Magn Reson **143**: 411
[35] Samoson A, Lippmaa E, Pines A (1988) Mol Phys **65**: 1013
[36] Llor A, Virlet J (1988) Chem Phys Lett **152**: 248
[37] Frydman L, Harwood J (1995) J Am Chem Soc **117**: 5367
[38] Medek A, Harwood JS, Frydman L (1995) J Am Chem Soc **117**: 12779
[39] McManus J, Kemp-Harper R, Wimperis S (1999) Chem Phys Lett **311**: 292
[40] Wi S, Frydman L (2000) J Chem Phys **112**: 3248
[41] Wi S, Frydman V, Frydman L (2001) J Chem Phys **114**: 8511
[42] Pike KJ, Malde RP, Ashbrook SE, McManus J, Wimperis S (2000) Solid State Nucl Magn Reson **16**: 203
[43] Jerschow A, Logan JW, Pines A (2001) J Magn Reson **149**: 268
[44] Marinelli L, Frydman L (1997) Chem Phys Lett **275**: 188
[45] Pruski M, Bailly A, Lang DP, Amoureux JP, Fernandez C (1999) Chem Phys Lett **307**: 35
[46] Gan Z (2000) J Am Chem Soc **122**: 3242
[47] Pike KJ, Ashbrook SE, Wimperis S (2001) Chem Phys Lett **345**: 400
[48] Ashbrook SE, Wimperis SJ (2002) J Magn Reson **156**: 269
[49] Ashbrook SE, Wimperis S (2002) ENC Conference Asilomar California, Poster M/T73
[50] Pines A, Gibby MG, Waugh JS (1973) J Chem Phys **59**: 569
[51] Stejskal EO, Schaefer J, Waugh JS (1977) J Magn Reson **28**: 105
[52] Schaefer J, Stejskal EO (1976) J Am Chem Soc **98**: 1031
[53] Hartmann SR, Hahn EL (1962) Phys Rev **128**: 2042
[54] Vega AJ (1992) Solid State Nucl Magn Reson **1**: 17

[55] Vega S (1981) Phys Rev A **23**: 3152
[56] Ashbrook SE, Wimperis S (2001) Chem Phys Lett **340**: 500
[57] Fernandez C, Delevoye L, Amoureux JP, Lang DP, Pruski M (1997) J Am Chem Soc **119**: 6858
[58] Ashbrook SE, Wimperis S (2000) Mol Phys **98**: 1
[59] Ashbrook SE, Brown SP, Wimperis S (1998) Chem Phys Lett **288**: 509
[60] Rovnyak D, Baldus M, Griffin RG (2000) J Magn Reson **142**: 145
[61] Lim KH, Grey CP (2000) J Chem Phys **112**: 7490
[62] Lim KH, Grey CP (1999) Chem Phys Lett **312**: 45
[63] Ashbrook SE, Wimperis S (2000) J Magn Reson **147**: 238
[64] De Paul SM, Ernst M, Shore JS, Stebbins JF, Pines A (1997) J Phys Chem B **101**: 3240
[65] Wang SH, De Paul SM, Bull LM (1997) J Magn Reson **125**: 364
[66] Steuernagel S (1998) Solid State Nucl Magn Reson **11**: 197
[67] Eastman MA (1999) J Magn Reson **139**: 98
[68] Chan JCC, Bertmer M, Eckert H (1998) Chem Phys Lett **292**: 154
[69] Chan JCC (1999) J Magn Reson **140**: 487
[70] Baldus M, Rovnyak D, Griffin RG (2000) J Chem Phys **112**: 5902
[71] Raleigh DP, Levitt MH, Griffin RG (1988) Chem Phys Lett **146**: 71
[72] Nijman M, Ernst M, Kentgens APM, Meier BH (2000) Mol Phys **98**: 161
[73] Gan Z, Robyr P (1998) Mol Phys **95**: 1143
[74] Walls J, Lim K, Pines A (2002) J Chem Phys **116**: 79
[75] Walls JD, Lim KH, Logan JW, Urban JT, Jerschow A, Pines A (2002) J Chem Phys **117**: 518
[76] Logan JW, Urban JT, Walls JD, Lim KH, Jerschow A, Pines A (2002) Solid State Nucl Magn Reson (in press)
[77] Wi S, Heise H, Logan J, Sakellariou D, Pines A (2002) ENC Conference Asilomar California, Poster W/Th66
[78] Hartmann P, Jäger C, Zwanziger JW (1999) Solid State Nucl Magn Reson **13**: 245
[79] Duer MJ, Painter AJ (1999) Chem Phys Lett **313**: 763
[80] Bax A, Freeman R, Frenkiel TA (1981) J Am Chem Soc **103**: 2102
[81] Wu G, Dong S (2001) Chem Phys Lett **334**: 265

Invited Review

Multiple-Quantum Magic-Angle Spinning: High-Resolution Solid State NMR Spectroscopy of Half-Integer Quadrupolar Nuclei

Amir Goldbourt[1] and **Perunthiruthy K. Madhu**[2,*]

[1] Department of Chemical Physics, Weizmann Institute of Science, Rehovot 76100, Israel
[2] Department of Chemistry, University of Southampton, Southampton SO17 1BJ, United Kingdom

Received April 16, 2002; accepted May 6, 2002
Published online September 19, 2002

Summary. Experimental and theoretical aspects of the multiple-quantum magic-angle spinning experiment (MQMAS) are discussed in this review. The significance of this experiment, introduced by *Frydman* and *Harwood*, is in its ability to provide high-resolution NMR spectra of half-integer quadrupolar nuclei ($I \geq 3/2$). This technique has proved to be useful in various systems ranging from inorganic materials to biological samples. This review addresses the development of various pulse schemes aimed at improving the signal-to-noise ratio and anisotropic lineshapes. Representative spectra are shown to underscore the importance and applications of the MQMAS experiment.

Keywords. NMR spectroscopy; Solid state; Quadrupolar nuclei; MQMAS; Amplitude modulation.

Introduction

The internal spin interactions in solid state nuclear magnetic resonance (NMR) that chiefly govern the spectral response are quadrupolar (Q), dipole–dipole (DD), chemical shift anisotropy (CSA), and scalar coupling (J). These interactions are in general anisotropic in character. The quadrupolar interaction vanishes in the case of spin-$\frac{1}{2}$ nuclei like ^{1}H and ^{13}C due to the internal spherical symmetry of the nuclear spin system. This is not the case for nuclei with spin quantum number $I \geq 1$. These nuclei possess an electric quadrupolar moment (also called nuclear quadrupole moment) in addition to having a nuclear magnetic moment as in the case of spin-$\frac{1}{2}$ nuclei. This nuclear quadrupole moment couples to the inhomogeneous

* Corresponding author. E-mail: madhu@soton.ac.uk

internal electric field gradients (EFG) [1]. The interaction of the nuclear quadrupole moment with a non-vanishing EFG at the nuclear site leads to a modification of the *Zeeman* energy levels and dominates the appearance of an NMR spectrum. The nature of EFG depends on the local electronic environment and symmetry about the nucleus. Hence, if the characteristics of the EFG can be determined from NMR experiments, information about the chemistry of a molecule, and details about the molecular or the crystalline environment can be ascertained.

Quadrupolar nuclei can be classified into those with integer spins (1, 3, etc.) and half-integer spins ($\frac{3}{2}$, $\frac{5}{2}$, etc.). Some prominent nuclei that fall into the class of spin-1 are deuterium (^{2}H) and nitrogen (^{14}N), spin-3 is boron (^{10}B), spin-$\frac{3}{2}$ are sodium (^{23}Na), rubidium (^{87}Rb), and boron (^{11}B), spin-$\frac{5}{2}$ are oxygen (^{17}O), aluminium (^{27}Al), and molybdenum (^{95}Mo), spin-$\frac{7}{2}$ are cobalt (^{59}Co), calcium (^{43}Ca), and scandium (^{45}Sc), and spin-$\frac{9}{2}$ is niobium (^{93}Nb). A survey of the periodic table shows that around 70% of the elements are quadrupolar in nature.

Several techniques are available to obtain a high-resolution spectrum of spin-$\frac{1}{2}$ nuclei by decoupling experiments and systematically probing the anisotropy of the interactions by recoupling experiments [2]. However, in samples containing quadrupolar nuclei, unless there is a spherical charge distribution around the nucleus, which occurs only in a few cases, the quadrupolar interaction far exceeds DD, CSA, and J interactions. Typically, the quadrupolar interactions vary from a few kHz to a few MHz leading up to 2000–3000 MHz in the case of ^{127}I.

Due to their abundance, quadrupolar nuclei are dominant in several molecular systems ranging from porous materials, ceramics and glasses, to superconductors and biological systems such as nucleic acids and proteins. The practical applications of quadrupolar nuclei in several branches of science were discussed a long time ago by *Pound* [3]. Two factors have impeded the routine study of these materials. One is the line broadening in a powder sample arising from the size of the quadrupole moment that causes lack of resolution in currently available spectrometers and secondly, the complexity in spin dynamics due to the presence of more than two energy levels. This renders development of pulse methodology a demanding task as against the scenario in spin-$\frac{1}{2}$. Utilising the established spin-$\frac{1}{2}$ techniques does not always provide the required result. For example, magic-angle-spinning (MAS) [4, 5] does not provide isotropic high-resolution spectra in quadrupolar systems and cross-polarisation (CP) to and from such nuclei are normally not efficient [6, 7]. Hence, although rich in information content (e.g. in ^{2}H systems), and sometimes enabling an easier interpretation of the spectrum, high-resolution quadrupolar split spectra in the solid state are difficult to acquire with conventional spin-$\frac{1}{2}$ means.

In order to overcome the resolution problems in quadrupolar systems, mechanical ways were first devised. This resulted in two ingenious attempts, namely, double rotation, DOR [8] and dynamic angle spinning, DAS [9, 10], both involving manipulation of the spatial part of the quadrupolar Hamiltonian. Recent attempts focussed on the manipulation of the spin part of the Hamiltonian and this paved the way to the introduction of MQMAS [11, 12] by *Frydman* and *Harwood*, and lately, satellite transition MAS (STMAS) by *Gan* [13]. The introduction of MQMAS led to a spurt of activities that saw a large number of applications and improvements of this scheme. The popularity of this technique as against its forerunners can be understood from the simplicity of the experimental protocol with no need for any extra hardware and demanding spectrometer performance efficiency. Development in this area has also shed more light on the complicated spin dynamics of quadrupolar nuclei.

In this review, we deal mainly with MQMAS, and refer to [14–16] for details on DOR, DAS, and STMAS. Many of the basic properties of the quadrupolar Hamiltonian have been dealt with in a succinct manner by *Vega* [17] and we borrow on them extensively here giving the appropriate equations wherever necessary.

The outline of this review is as follows: Introduction to the quadrupolar Hamiltonian under MAS, the technique of MQMAS and interpretation of MQMAS spectra, possible solutions to the problem of sensitivity, some heteronuclear techniques, and finally a few representative applications and future perspectives.

Quadrupolar Spins – Theory and Methods

The quadrupolar Hamiltonian, $\mathcal{H}_Q$, besides being stronger than the other internal spin Hamiltonians, is often stronger than typical radiofrequency (rf) fields employed in NMR. For an analysis of quadrupolar spectra, the first two terms in the expansion of $\mathcal{H}_Q$ are considered: the first-order, $\mathcal{H}_Q^{(1)}$, and second-order, $\mathcal{H}_Q^{(2)}$, following standard perturbation theory.

Figure 1 shows the energy levels of spins $I = \frac{3}{2}$ and 1 in a *Zeeman* field ν_0 in the absence and presence of a quadrupolar interaction ν_Q. Also shown is a schematic of the resultant quadrupolar split NMR spectra for the case of a single crystal. The quadrupolar interaction shifts the eigenvalues of the *Zeeman* Hamiltonian, resulting in an NMR spectrum that is split into $2I$ peaks. In a powdered sample, lines are broadened by the orientation dependence of the first- and second-order quadrupolar interactions.

The first-order quadrupolar interaction affects only the satellite transitions $m \leftrightarrow m \pm 1$, where $|m|$ takes the values $\frac{1}{2}, \frac{3}{2}, \frac{5}{2}, \cdots$. Thus, in the case of integer spins, the observable satellite transitions are broadened extensively (up to several MHz), making their detection cumbersome (an exception is deuterium, ^{2}H, which possesses a relatively small quadrupolar coupling constant (≈ 200 kHz), hence being a commonly probed nucleus). The broadening is a result of the significant shift the energy levels experience due to $\mathcal{H}_Q^{(1)}$. However, the symmetric transitions $m \leftrightarrow -m$ are affected only by the relatively smaller second-order quadrupolar interaction. Hence, the observable single-quantum transition of half-integer quadrupolar spins, $-\frac{1}{2} \leftrightarrow \frac{1}{2}$, is broadened only by $\mathcal{H}_Q^{(2)}$. This transition, called the central transition, thus provides a relatively narrow powder lineshape, which retains all information about the nuclear quadrupole coupling constant, NQCC, and the asymmetry of the quadrupolar interaction tensor, η.

NQCC characterises the size of the quadrupolar interaction experienced by a particular nucleus and is given in frequency units by e^2qQ/h, denoted by χ. Here, eq is the magnitude of the EFG and eQ is the nuclear electric quadrupole moment in units of the electronic charge e [18]. The quadrupolar frequency ν_Q in units of Hz or ω_Q in units of rad s^{-1}, is given by [1],

$$\nu_Q = \frac{\omega_Q}{2\pi} = \frac{3\chi}{2I(2I-1)}. \tag{1}$$

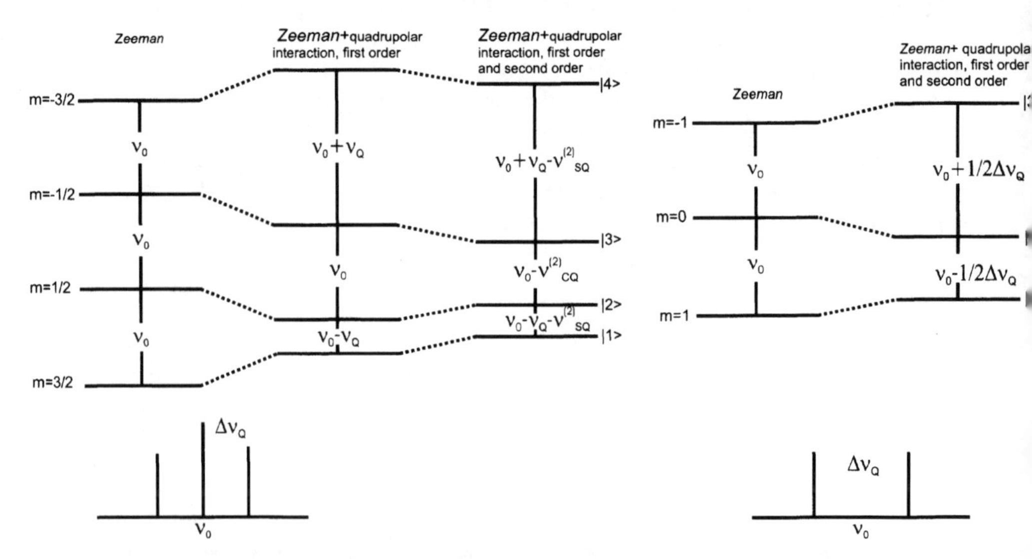

Fig. 1. Energy levels of a spin-$\frac{3}{2}$ and a spin-1 quadrupolar system in the presence of *Zeeman* field, first-order quadrupolar and second-order quadrupolar interactions. The allowed transitions, $\Delta m = \pm 1$, are indicated. The stick spectra at the bottom are the quadrupolar split spectra for a single crystal. ν_Q is the first-order quadrupolar frequency, $\nu_{CQ}^{(2)}$ and $\nu_{SQ}^{(2)}$ are the second-order quadrupolar frequencies of central and satellite transitions respectively, $\omega_0 = 2\pi\nu_0$ is the *Larmor* frequency, and $\Delta\omega_Q = 2\pi\Delta\nu_Q$ is the quadrupolar splitting

Hence, for a spin-$\frac{3}{2}$, for example, $\nu_Q = \frac{1}{2}\chi$, and for a spin-$\frac{5}{2}$, $\nu_Q = \frac{3}{20}\chi$. The NQCC vanishes for those quadrupolar nuclei positioned at a cubic site due to the inherent symmetry. The second-order quadrupolar interactions scale as $\frac{\nu_Q^2}{\nu_0}$. Hence, a higher resolution is obtained at higher magnetic fields for the same quadrupolar strength.

The MAS Hamiltonian

The general form of the quadrupolar Hamiltonian in the notation of irreducible tensor operators [1, 17] is given in Eq. (2) as

$$\mathcal{H} = \frac{\nu_Q}{3} \sum_{k=-2}^{2} (-1)^k T_q^{(2)} V_{-q}^{(2)} \tag{2}$$

with

$$\begin{aligned} T_0^{(2)} &= \sqrt{\frac{1}{6}}[3I_z^2 - I(I+1)] \\ T_{\pm 1}^{(2)} &= I_z I_\pm = I_\pm I_z \\ T_{\pm 2}^{(2)} &= I_\pm^2 . \end{aligned} \tag{3}$$

The elements $V_q^{(2)}$ are obtained from the components of the quadrupolar tensor $\rho_m^{(2)}$ in their principal axis system (PAS) by a transformation to the laboratory frame using the *Euler* angles α, β, γ:

$$V_q^{(k)} = \sum_{q'=-k}^{k} \rho_{q'}^{(2)} D_{q'q}^{(2)}(\alpha, \beta, \gamma). \tag{4}$$

The terms $D_{q'q}^{(k)}(\alpha, \beta, \gamma)$ are the *Wigner* matrix elements [19], and

$$\begin{aligned} \rho_0^{(2)} &= \sqrt{\frac{3}{2}} \\ \rho_{\pm 1}^{(2)} &= 0 \\ \rho_{\pm 2}^{(2)} &= -\frac{1}{2}\eta. \end{aligned} \tag{5}$$

In the laboratory frame the quadrupolar interaction can be treated as a perturbation to the *Zeeman* field, and can be expressed as a sum of first-order and second-order terms, as follows:

$$\begin{aligned} \mathcal{H}_Q &= H_Q^{(1)} + H_Q^{(2)} \\ H_Q^{(1)} &= \frac{\nu_Q}{3} T_0^{(2)} V_0^{(2)} = \frac{h\nu'_Q}{6}[3I_z^2 - I(I+1)] \\ H_Q^{(2)} &= \frac{h\nu_Q^2}{9\nu_0}\left\{2I_z\left[2I_z^2 - I(I+1) + \frac{1}{4}\right]V_{-1}^{(2)}V_1^{(2)}\right. \\ &\qquad \left. + I_z\left[I_z^2 - I(I+1) + \frac{1}{2}\right]V_{-2}^{(2)}V_2^{(2)}\right\} \end{aligned} \tag{6}$$

where ν'_Q is given by

$$\nu'_Q = \nu_Q\left(\frac{3\cos^2\beta - 1}{2} + \frac{\eta}{2}\sin^2\beta\cos 2\alpha\right). \tag{7}$$

Under MAS conditions, the PAS of the quadrupolar spin has to be first transformed into the rotor frame with the proper *Wigner* matrix $D(\alpha, \beta, \gamma)$. The angles (α, β, γ) describe the location of the rotor axis in the PAS of the quadrupolar tensor. Only then the transformation into the LAB frame is performed using $D(\omega_r t, \theta_{MA}, 0)$, $\omega_r t$ being the phase angle experienced by the rotor while spinning at the magic angle, θ_{MA} (54.7°) with respect to the static magnetic field. Thus Eq. (6) describes the MAS average Hamiltonian for the quadrupolar nucleus with a new definition of the terms $V_q^{(2)}$:

$$V_q^{(2)} = \sum_{m=-2}^{2} D_{mq}^{(2)}(\omega_r t, \theta_{MA}, 0) \sum_{n=-2}^{2} D_{nm}^{(2)}(\alpha, \beta, \gamma)\rho_n^{(2)}. \tag{8}$$

The terms $V_q^{(2)} V_{-q}^{(2)}$ in Eq. (6) contain products of *Wigner* matrix elements. These may be expanded using the *Clebsch-Gordon* coefficients [19]. Hence, in the fast spinning limit, the spatial part of the quadrupolar second-order Hamiltonian becomes a sum of *Legendre* polynomials with ranks 0, 2, and 4 (the terms with ranks 1 and 3 vanish due to symmetry). After such an expansion, a straightforward calculation gives a simple expression for the symmetric transitions,

$$\nu_{m,-m} = \sum_{l=0,2,4} \nu_Q^{(l)}(\alpha, \beta, \gamma) C_l(I, m) P_l(\cos \theta_{MA}). \tag{9}$$

In Eq. (9) $P_l(\cos \theta)$ is the *Legendre* polynomial of rank l, and $C_l(I, m)$ are zero-, second- and fourth-rank (spin) coefficients depending on the spin quantum number I and order $2m$ of the transition. Their explicit forms may be found in Ref. [11] and are reproduced here:

$$\begin{aligned} C_0(I, m) &= 2m[I(I+1) - 3m^2] \\ C_2(I, m) &= 2m[8I(I+1) - 12m^2 - 3] \\ C_4(I, m) &= 2m[18I(I+1) - 34m^2 - 5]. \end{aligned} \tag{10}$$

Equation (9) is the basis of the MQMAS experiment as will be discussed later on. Additional details about the functional forms of lineshapes and shifts of quadrupolar nuclei can be found in Ref. [20, 21].

Magic-Angle Spinning Spectra

As outlined above, the central transition of non-integer quadrupolar spins is not affected by the first-order quadrupolar interaction. Hence, it is amenable to NMR spectroscopy using standard NMR techniques. Nevertheless, the existence of several energy levels complicates the excitation profile of a quadrupolar spin. While for a spin-$\frac{1}{2}$ system the nutation frequency ν_{nut} equals the rf field strength $\nu_1(=\gamma B_1)$, for a quadrupolar spin I, $\nu_{nut} = \left(I + \frac{1}{2}\right)\nu_1$. This equation holds goods as long as the excitation is selective, that is $\nu_1 \ll \nu_Q$. A hard-pulse excitation ($\nu_1 \gg \nu_Q$) gives a nutation frequency that is similar to a spin-$\frac{1}{2}$, $\nu_{nut} = \nu_1$. In the intermediate regime, a complex behaviour is expected and the nutation frequency of the central transition depends on ν_Q. Thus, in order to obtain a quantitative MAS spectrum, a small angle pulse $\left(\nu_1 t \leq \frac{\pi}{6}\right)$ has to be used [22, 23]. This ensures that in a sample of multiple sites with varying values of ν_Q, each site is uniformally excited. Figure 2 illustrates this aspect for a spin-$\frac{3}{2}$ nucleus.

Theoretical second-order quadrupolar broadened line shapes are given in Fig. 3 for both static and MAS cases for a few values of η. Although MAS achieves considerable line narrowing of the central transition (3 to 4 times with respect to the line width in a static case) it is still beyond what can be called a high-resolution quadrupolar spectrum.

The reason why MAS cannot narrow the second-order broadening can be seen directly from Eq. (9). The isotropic quadrupolar shift ($l = 0$) is given explicitly in Eq. (11) as

$$\nu_{iso}^{(2)} = \frac{1}{30} \frac{\nu_Q^2}{\nu_0} \left[I(I+1) - \frac{3}{4} \right] \left(1 + \frac{1}{3}\eta^2 \right). \tag{11}$$

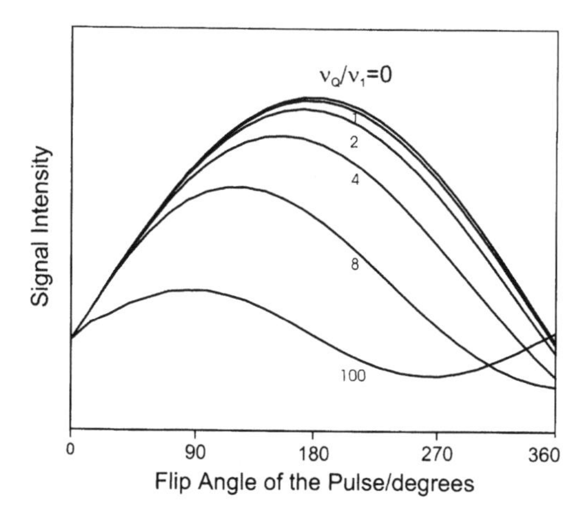

Fig. 2. The single-quantum signal intensity (normal MAS spectral intensity with a single pulse for excitation) plotted in arbitrary units as a function of the flip angle of the excitation pulse. The curves are calculated for different values of the ratio of the ^{23}Na quadrupolar frequency, ν_Q, to the rf field strength, $\nu_1 = 100.0\,\text{kHz}$, as indicated against each curve. The MAS rate was 10.0 kHz and the ^{23}Na *Larmor* frequency was 91.25 MHz at a magnetic field of 8.1 T. This figure illustrates that the excitation of a powder sample is non-uniform, unless a small flip angle pulse is used

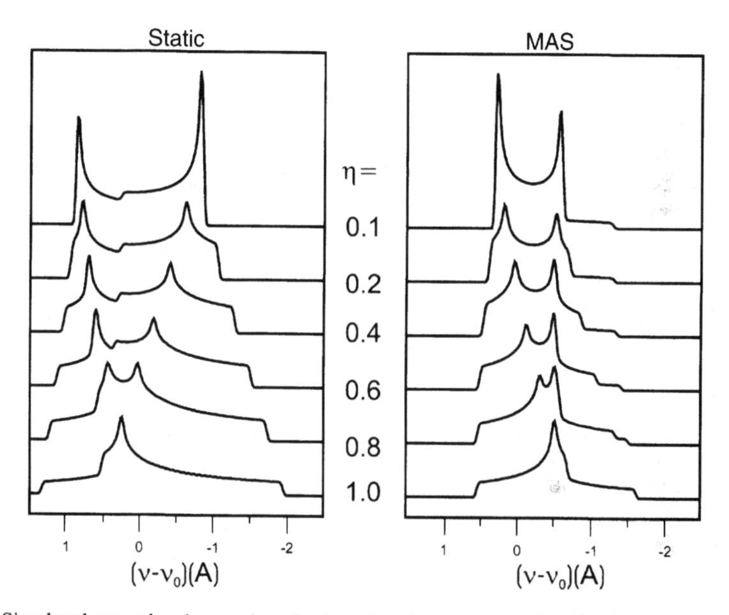

Fig. 3. Simulated second-order quadrupolar broadened central transition lineshapes of half-integer quadrupolar spins for both static and MAS conditions. The frequency scale is in units of A, given by $A = \frac{1}{9}\left[I(I+1) - \frac{3}{4}\right]\frac{\nu_Q^2}{\nu_0}$. Note that A is a universal parameter and, hence, the lineshapes do not depend on I, but the widths do

The product of the orientation dependent terms in Eq. (9) with the second- and fourth-rank *Legendre* polynomials, given in Eq. (12),

$$P_2(\cos\theta) = \frac{1}{2}(3\cos^2\theta - 1)$$
$$P_4(\cos\theta) = \frac{1}{8}(35\cos^4\theta - 30\cos^2\theta + 3) \tag{12}$$

is the source for the line broadening of the central transition spectra, where θ is the angle of rotation axis with respect to the static external magnetic filed. It is evident from these equations that there are no common roots for both of them. Hence, a rotor spinning at the magic-angle, $\theta_{MA} = arc\tan 1/\sqrt{2}$, will zero the $P_2(\cos\theta)$ terms, but will always leave behind an anisotropic contribution from the $P_4(\cos\theta)$ terms that have roots at 30.56° and 70.12°. This anisotropic contribution gives rise to the partial broadening of the MAS lineshapes as depicted in Fig. 3.

While only the nutation behaviour of the central transition was discussed above, it should be noted that other transitions are also possible with $\Delta m = \pm 1$ (satellite transitions) and more importantly, those with $\Delta m = \pm 3, \pm 5, \ldots$ *i.e.* the symmetric transitions. The excitation of the multiple quantum symmetric transitions scales as $\nu_{MQ} \propto \frac{\nu_1^M}{\nu_Q^{M-1}}$, where M is the order of the excited coherence [24]. The possibility to directly excite these multiple-quantum transitions directly is also crucial for the success of the MQMAS experiment.

Resolution Enhancement Schemes

Before introducing MQMAS, we outline here a few experiments, all aimed to probe quadrupolar systems to obtain the relevant quadrupolar parameters. In addition to MAS, these include nutation spectroscopy [25], variable angle spinning (VAS) [26], and satellite transition spectroscopy [27]. Nutation spectroscopy and VAS allow determination of ν_Q and η from the second-order quadrupolar line shape of the central transition. Their applicability is limited to instances where the second-order quadrupolar line shapes can be readily resolved. Satellite transition spectroscopy gives quadrupolar parameters by a simulation of the complete manifold of spinning sidebands for the satellite transitions in the MAS NMR spectra of quadrupolar nuclei. This is possible, since, the broadening of the sidebands of the satellite transitions is smaller than those of the central transition. However, the manifold of sidebands extends over several MHz and a very accurate setting of the magic-angle is required. Yet, none of these methods achieves real high-resolution spectra since an inhomogeneous contribution to the quadrupolar interaction still remains.

The two experimental approaches that yielded "real" high-resolution spectra for the first time were double rotation (DOR) [8] and dynamic angle spinning (DAS) [9, 10]. DOR involves two rotors, an outer rotor, inclined at an angle of θ_1 with respect to the external magnetic field, and an inner rotor containing the sample having an axis of rotation that makes an angle of θ_2 with the outer rotor. The DOR condition can be stated as, Eq. (13),

$$P_2(\cos\theta_1)P_2(\cos\theta_2) = 0$$
$$P_4(\cos\theta_1)P_4(\cos\theta_2) = 0, \tag{13}$$

with their solutions, Eq. (14), being either

$$\{\theta_1, \theta_2\} = \{54.73°, 30.56°\} \quad \text{or}$$
$$\{\theta_1, \theta_2\} = \{54.73°, 70.12°\}. \tag{14}$$

The DAS scheme makes use of only one rotor, but the orientation of the rotor axis with respect to the external magnetic field toggles between two angles for two equal periods of time. It is a two-dimensional (2D) experiment in which the rotor is spun for a time t_1 at an angle θ_1, then the magnetisation is stored along the z-axis while the spinner angle is hopped ($\approx 30\,\text{ms}$), and finally coherence is transferred back to the central transition with rotor spinning at an angle θ_2. The two rotor axis angles are chosen in such a way that the conditions, Eq. (15),

$$P_2(\cos\theta_1) + P_2(\cos\theta_2) = 0$$
$$P_4(\cos\theta_1) + P_4(\cos\theta_2) = 0 \tag{15}$$

are satisfied. A simple DAS solution is $\{\theta_1, \theta_2\} = \{37.38°, 79.19°\}$.

Both DOR and DAS are technically very demanding and are limited by several factors. In the DAS experiment the T_1 relaxation time of the quadrupolar nuclei has to be larger than the hopping duration of the rotor. Spin exchange mediated by dipolar interaction can occur during the relatively long hopping time and an extra broadening might appear from large chemical shift anisotropies, since they are not completely averaged out. The DOR experiment is limited by the relatively slow spinning rates available thus producing a complex manifold of sidebands.

By manipulating the spin part of the Hamiltonian during MAS, *Frydman* and *Harwood* showed that a high-resolution NMR spectrum can be obtained by correlating multiple and single quantum coherences [11]. The technique was named multiple-quantum magic-angle spinning, MQMAS, and was introduced in 1995. Recently, using similar ideas, another method based on spin averaging was proposed by *Gan* called satellite transition MAS (STMAS) [13, 28]. In this experiment the single-quantum satellite transitions rather than the multiple-quantum transitions are correlated with the central transition. The main limitations of this technique seem to be the requirement for a very accurate setting of the magic angle (down to 0.1%) and the need for synchronised t_1 acquisition, which limits the available spectral widths. These issues are currently being investigated.

The rest of this review deals with MQMAS and its development. The ease in its implementation compared with DOR and DAS can be gauged from the numerous publications and applications that have resulted ever since its inception.

MQMAS-Principle and Practical Implementation

MQMAS is a 2D solid state MAS NMR experiment. It results in a high-resolution spectrum along the indirect dimension, correlated with its corresponding MAS spectrum in the detection dimension. This is achieved by simultaneously manipulating spin and spatial parts of the quadrupolar Hamiltonian. MAS takes the role of spatial averaging and removes the chemical shift anisotropy, CSA, the heteronuclear dipolar interactions, and the second rank elements of the second-order quadrupolar interaction. Radiofrequency pulses are used to manipulate the spin

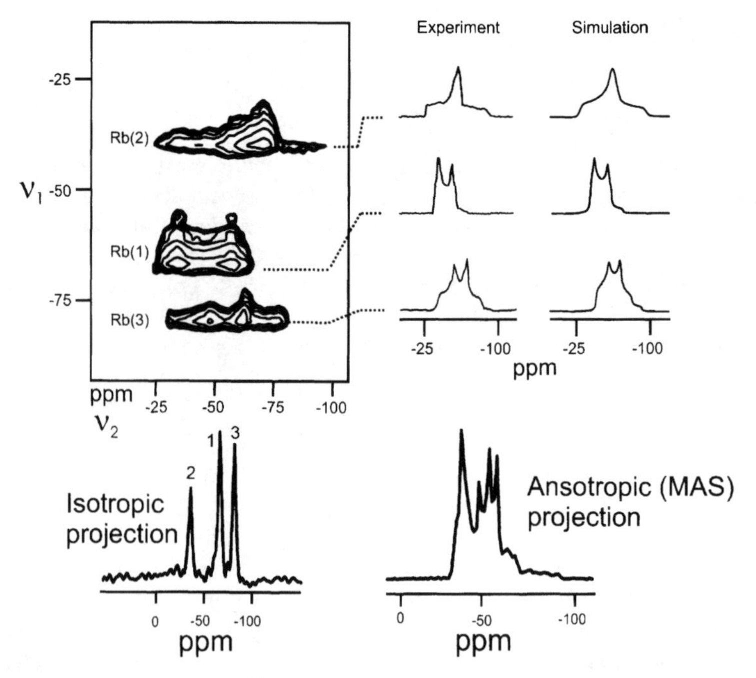

Fig. 4. Two-dimensional 3QMAS ^{87}Rb spectrum of $RbNO_3$ together with anisotropic slices corresponding to the three Rb sites from both experiment and simulations. Also shown are the isotropic and anisotropic projections, the former being the high-resolution quadrupolar spectrum and the latter corresponding to the normal MAS spectrum. The spectrum was recorded at a magnetic field of 4.7 T, MAS rate of 10.0 kHz, and rf strength of 70.0 kHz. The Rb sites are assigned both in the 2D spectrum and in the isotropic projection

part and average out the fourth rank elements of the second-order quadrupolar Hamiltonian. An echo is formed by correlating the frequencies of symmetric multiple-quantum coherences (MQC) and those of the central transition coherences (single-quantum coherence, SQC) for all crystallites in the powder simultaneously. Figure 4 shows a 2D ^{87}Rb MQMAS spectrum of a sample of polycrystalline rubidium nitrate. The isotropic projection (t_1 in time domain and F_1 or ν_1 in the frequency domain) clearly shows the three resolved ^{87}Rb sites. Sum projections on to the MAS dimension (t_2) of the three sites along with their corresponding simulations (theoretical lineshapes) demonstrate the possibility for obtaining the quadrupolar parameters.

Theory

Using Eq. (9), the evolution undergone by a $-m \leftrightarrow +m$ spin coherence can be expressed in terms of its accumulated phase ϕ given in Eq. (16) as [11, 12, 33]

$$\begin{aligned}\phi(m,\theta,\alpha,\beta,t) = \nu^{CS}2mt + \nu_Q^{(0)}C_0(I,m)t + \nu_Q^{(2)}(\alpha,\beta)C_2(I,m)P_2(\cos\theta)t \\ + \nu_Q^{(4)}(\alpha,\beta)C_4(I,m)P_4(\cos\theta)t\end{aligned} \quad (16)$$

where θ is the angle of rotation axis with respect to the static magnetic field, B_0. In the above ν^{CS} is associated with isotropic chemical shift. The C coefficients that depend on the spin quantum number I and the order of excited coherence, $M=2m$, are given in Eq. (10). In a 2D experiment, spin coherences evolve during times t_1 and t_2. In order to obtain an isotropic echo signal, the anisotropic part of the phase ϕ should be set to zero. This can be done with the constraints in Eq. (17):

$$\nu_Q^{(2)}(\alpha,\beta)C_2(I,m_1)P_2(\cos\theta_1)t_1 + \nu_Q^{(2)}(\alpha,\beta)C_2(I,m_2)P_2(\cos\theta_2)t_2 = 0$$
$$\nu_Q^{(4)}(\alpha,\beta)C_4(I,m_1)P_4\cos(\theta_1)t_1 + \nu_Q^{(4)}(\alpha,\beta)C_4(I,m_2)P_4\cos(\theta_2)t_2 = 0. \quad (17)$$

By setting $m_{1,2}=\frac{1}{2}$ (a single-quantum experiment) and the echo position to $t_2=t_1$ all terms but those of Eq. (15) are eliminated, and the conditions for the DAS experiment are met. An alternative way to accomplish the same task is by manipulating the spin coherences, so that, by setting $\theta_1=\theta_2=\theta_{MA}$ and leaving $m_{1,2}$ as parameters, the above constraints reduce to Eq. (18) given by,

$$C_4(I,m_1)t_1 + C_4(I,m_2)t_2 = 0. \quad (18)$$

Under these conditions, all 2nd rank anisotropies are removed, and since the expression above is independent of $\nu_Q^{(4)}$, all crystallites are refocused at a time $t_2 = -\frac{C_4(I,m_1)}{C_4(I,m_2)}t_1$ giving rise to a quadrupolar echo. This is the idea behind the MQMAS experiment.

In a MQMAS experiment, any order of MQC $2m_i$ can be used keeping in mind that the detection must be performed at the central transition. Thus, the experiment is performed by exciting MQC using a rf pulse and then converting them into single-quantum observable coherences using another rf pulse. However, recently *Jerschow et al.* showed that for a half-integer spin with $I>\frac{3}{2}$, higher order coherences can be correlated within each other, e.g. five-quantum (5Q) and triple-quantum (3Q) coherences, as long as an additional detection pulse is employed at the end of the sequence [34]. The MQ echo is picked up using a proper phase cycle and thus only the desired coherence pathway is selected. While experiments correlating MQ and SQ coherences are named 3QMAS, 5QMAS etc., the ones correlating various MQC's are named MQ/NQ MAS, where M and N are any two higher coherences. In principle, the highest resolution is attained by choosing the highest possible coherences [34, 35]. Thus, for a spin $\frac{5}{2}$, 3QMAS, 5QMAS and 5Q/3QMAS can all be performed with 5Q/3QMAS having the highest resolution.

Routinely, MQMAS is performed by correlating MQC with SQC giving rise to a purely isotropic second-order quadrupolar echo at a time

$$t_2 = \left|C_4(I,m_1)/C_4\left(I,\frac{1}{2}\right)\right|t_1. \quad (19)$$

Since the ratio $\frac{C_4(I,m_1)}{C_4(I,\frac{1}{2})}$ can be of any sign, additional spin manipulation has to be performed in order to observe the echo, as will be discussed in the next section. A signal arising from a positive value of the ratio is called an echo and that arising from a negative value is called an antiecho. Table 1 gives the value of $k=\left|C_4(I,m_1)/C_4\left(I,\frac{1}{2}\right)\right|$ for various I and m values. Extension of this table to the MQ/NQ experiment is straightforward using the $C_l(I,m)$ coefficients in Eq. (10).

Table 1. The values of $k(I, m)$ in MQMAS

m/I	3Q	5Q	7Q	9Q
$\frac{3}{2}$	$\frac{7}{9}$			
$\frac{5}{2}$	$\frac{19}{12}$	$\frac{25}{12}$		
$\frac{7}{2}$	$\frac{101}{45}$	$\frac{11}{9}$	$\frac{161}{45}$	
$\frac{9}{2}$	$\frac{91}{36}$	$\frac{95}{36}$	$\frac{7}{18}$	$\frac{31}{6}$

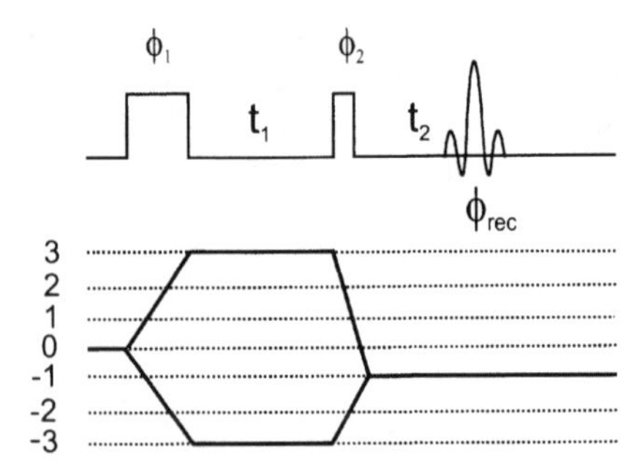

Fig. 5. The basic two-pulse (CW–CW) sequence used for MQMAS experiments along with the coherence pathways selected for 3Q experiments. The required phase cycle can be explicitly written as, $\phi_1 = 0°, 60°, 120°, 180°, 240°, 300°$, $\phi_2 = 0°$ and $\phi_{rec} = 0°, 180°$. (The same sequence may be used for 5QMAS with the following phase cycling: $\phi_1 = 0°, 36°, 72°, 108°, 144°, 180°, 216°, 252°, 288°, 324°$, $\phi_2 = 0°$ and $\phi_{rec} = 0°, 180°$. This phase cycling will select $0 \rightarrow \pm 5 \rightarrow -1$ pathways)

Pulse Schemes

The first MQMAS sequence as suggested by *Frydman* and *Harwood* consisted of three pulses, analogous to liquid state NMR sequences such as DQF-COSY and INADEQUATE [36, 37]. Based on the work of *Vega* and *Naor* on half-integer quadrupolar system (^{23}Na) [24] and *Vega* and *Pines* on double-quantum deuterium NMR [38], a two-pulse sequence was suggested by several authors independently [12, 39, 40]. The two-pulse scheme, Fig. 5, served as a foundation for subsequent developments in MQMAS. The first pulse excites all possible MQ coherences and the desired one is chosen by an appropriate phase cycling. After the MQ evolution the second pulse converts the MQC to 1QC resulting in an echo. In the two pulse scheme, the t_1 evolution time in the 2D experiment is governed solely by MQC evolution. Since the echo is formed at times $t_2 = kt_1$, it is tilted in the k direction leading to a time domain set as shown in Fig. 6a. Such a data set needs to be sheared to get a pure isotropic 2D *Fourier* transformed (FT) spectrum. The shearing is performed by employing a linear phase correction in the mixed (t_1, ν_2) domain [41]. The results of such a procedure is shown in Fig. 6(b–d). In order to obtain pure absorption 2D lineshapes, some modifications to the original sequence are needed and the three common methods are described in the following.

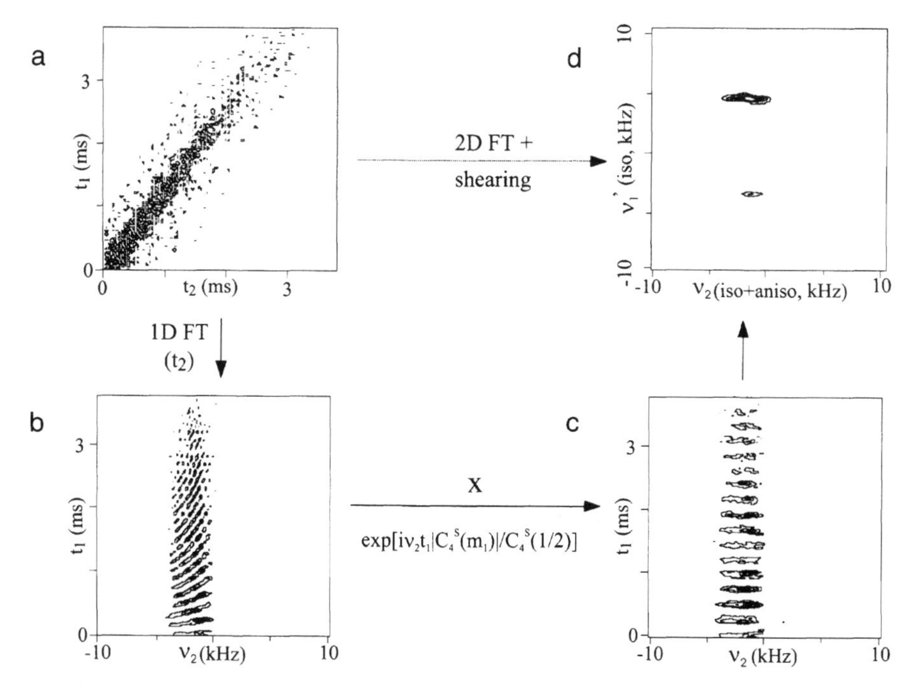

Fig. 6. Schematics showing the 2D time domain, shearing, and frequency domain representation of MQMAS spectrum: (a) 3Q → 1Q correlation experiment obtained using the pulse sequence shown in Fig. 5. Note that the free induction decay, fid, has a slope of $\frac{7}{9}$ (Table 1). The shearing transformation is illustrated in (b) and (c) which is done via a first-order t_1-dependent phase correction along the F_2 axis (ν_2). (d) Isotropic → anisotropic correlation spectrum obtained after a *Fourier* and shearing transformation of the time-domain data set in (a). The spectrum corresponds to ^{23}Na 3QMAS of sodium oxalate, $Na_2C_2O_4$. The prime in (d) on ν_1 denotes the result after the shearing transformation. (Reproduced with permission from Ref. [12], Copyright (1995), American Chemical Society)

z-Filter Experiments

The *z*-filter was adapted to MQMAS from liquid state NMR [36] by *Amoureux et al.* [42]. This three pulse scheme (Fig. 7a) consists of a MQ excitation pulse, a mixing pulse that transfers $\pm m$ coherences into 0Q coherence, thereby introducing a *z*-filter and finally a soft 90° pulse that creates observable magnetisation ($m = -1$ coherence). The transfer of $(+m) \rightarrow (-1)$ and $(-m) \rightarrow (-1)$ does not occur with the same efficiency in the two-pulse experiment, Fig. 5. The middle pulse in the *z*-filter experiment forces an equal transfer by performing $\pm m \rightarrow 0$ coherence transfer. This scheme ensures a pure cosine modulation along the indirect dimension, t_1. Phase modulated data set in t_1 can be obtained by using hypercomplex acquisition or by employing TPPI/STATES [36]. The frequency spectrum is obtained by *Fourier* transforming along the direct dimension, F_2, performing a shearing transformation with a phase correction of $e^{ik\omega_2 t_1}$ of all points and again *Fourier* transforming along F_1, as demonstrated in Fig. 6.

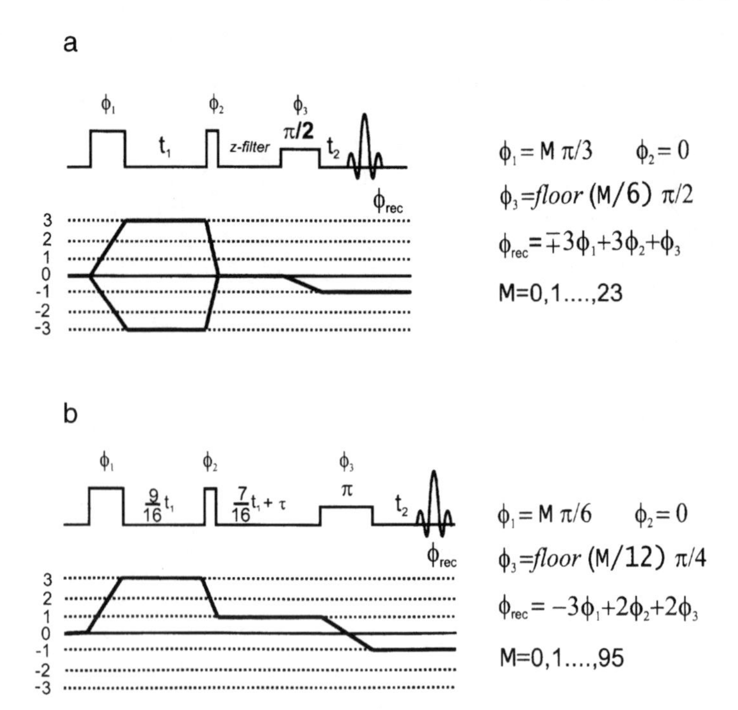

Fig. 7. MQMAS pulse sequences using (a) z-filter and (b) split-t_1 whole-echo scheme (this form of split-t_1 needs to be used for 3QMAS in spin-$\frac{3}{2}$) providing pure phase absorptive spectra with (b) avoiding the need for a shearing transformation. The phase cycle is mentioned in the figure. *floor*(x) returns the largest integer $\leq x$. In general, split-t_1 experiments have delays $\frac{1}{1+k}t_1$ and $\frac{k}{1+k}t_1$ which for spin-$\frac{3}{2}$ become $\frac{9}{16}t_1$ and $\frac{7}{16}t_1$

Phase-Modulated Split-t_1 Whole-Echo

Whole-echo acquisition [43], adapted to MQMAS by *Massiot et al.* [40], is an alternative way to obtain pure phase spectra. It relies on the fact that an echo of the fid can be induced by a π pulse without significant signal loss. *Fourier* transformation of such an echo signal leads to pure absorptive spectrum with a vanishing dispersive part, as long as relaxation effects are negligible.

The split-t_1 method, introduced by *Brown et al.* [44], gets rid of the need to do a shearing transformation by combining the MQ and 1Q (or NQ) evolution periods in the t_1 time domain. A combination with the whole-echo acquisition method results in a pure phase 2D MQMAS spectra, with ridges lying parallel to the F_2 frequency axis [45]. Figure 7b shows split-t_1 whole-echo pulse sequence for spin-$\frac{3}{2}$ where 3QC and 1QC evolve for times $\frac{9}{16}t_1$ and $\frac{7}{16}t_1$ respectively. These values correspond to the general required delays of $\frac{1}{1+k}$ for MQ evolution and $\frac{k}{1+k}$ for 1Q evolution.

The pulse sequence consists of three pulses: an excitation pulse, a conversion pulse and a soft π pulse to shift the echo. There are two different versions of this experiment, depending on the sign of $C_4(I, m_1)/C_4(I, \frac{1}{2})$ (see Fig. 13). Since the echo always appears at a constant time during t_2, τ_{echo}, this time should be chosen such that the whole echo appears in the fid. Signal processing includes: (1) putting the position of the echo signals at the middle of the fid's and zero filling symmetrically; a *Gauss*ian weighted window can be applied to the time domain data if desired, (2) redefining the time origin to be zero at the centre of the echo (this is similar to swapping the two halves of the spectrum), and (3) complex 2D FT (including apodisation, phasing etc.). An alternative approach includes (1) FT along t_1, (2) inversion of alternating points, and (3) FT along t_2.

Hypercomplex Experiments

Another method to obtain pure absorption line shapes is to collect a set of hypercomplex data. Two experiments are performed consecutively with a phase shift of $\frac{90^\circ}{M}$ on the first pulse, M being the order of coherence evolving during t_1. The resulting two signals, S_x and S_y, can be combined to yield echo and anti-echo signals by the following linear combinations:

$$\begin{aligned} S'_e(t_1, t_2) &= S_x - iS_y \\ S'_a(t_1, t_2) &= S_x + iS_y. \end{aligned} \tag{20}$$

A shearing transformation can now be applied after FT along F_2 (if needed):

$$\begin{aligned} S_e(t_1, w_2) &= e^{ikw_2t_1} S'_e(t_1, w_2) \\ S_a(t_1, w_2) &= e^{-ikw_2t_1} S'_a(t_1, w_2). \end{aligned} \tag{21}$$

The FT will create two sets of 2D signals which can be combined according to

$$S(w_1, w_2) = S_e(w_1, w_2) + S_a(-w_1, w_2) \tag{22}$$

thus giving rise to pure absorption mode lineshapes.

Additional Pulse Scheme Combinations

In addition to the three schemes mentioned above, several other pulse combinations may be considered. A z-filter sequence can be combined with split-t_1 experiments by using the following coherence pathway: $\pm 3 \rightarrow \pm 1 \rightarrow 0 \rightarrow -1$. This makes use of 4 rf pulses, the first two hard rf pulses for excitation of MQC and their conversion to ± 1 coherences. The latter two are soft $\pi/2$ pulses. Another option is to perform a split-t_1 experiment without shifting the echo. In this case a hypercomplex acquisition should be performed, without the need for a shearing transformation. Sometimes such combinations can be advantageous, since sensitivity enhancement schemes, which will be discussed later, can be successfully applied to them.

Data Analysis – Extraction of NMR Parameters

After applying the *Fourier* transformation, the axes have to be properly labelled and referenced according to the experimental conditions. The labelling scheme corresponds to either the split-t_1 frequency domain data, or to a sheared MQMAS spectrum. This labelling is not straightforward since the evolution in the F_1 domain is governed by isotropic chemical and second-order quadrupolar shifts.

There are several different ways in the literature to label the axes [46, 47]. We adopt here a method in which the apparent *Larmor* frequency does not scale with M

(the order of MQ coherence evolving during t_1), but its sign can change according to the specific experiment. A negative *Larmor* frequency is used (*i.e.* the isotropic axis is inverted before any referencing is performed) if $M<2I$, and positive otherwise. Referencing the spectrum along the F_1 dimension should take into account the evolution of the isotropic chemical shift (δ_{cs}) under MQ and 1Q coherences. The reference frequency in F_1 is defined therefore as $\frac{k-M}{1+k}\delta$, δ being the shift in ppm with respect to a reference sample. This expression is the sum of $\frac{-M}{1+k}\delta$, chemical shift resulting from the evolution of MQC's, and $\frac{k}{1+k}\delta$, chemical shift resulting from the evolution of single-quantum coherence. The expressions for δ_1 (frequency shift in ppm along F_1) and δ_2 (centre of gravity of peaks along F_2) depend on δ_{cs} and on the second order quadrupole shift (δ_Q) and are given by Eq. (23) and Eq. (24) as [48],

$$\delta_1 = \frac{-p+k}{1+k}\delta_{cs} + \frac{A(I,p) - kA(I,-1)}{1+k}\delta_Q \tag{23}$$

$$\delta_2 = \delta_{iso} + A(I,-1)\delta_Q \tag{24}$$

where

$$A(I,p) = \frac{p}{30}(4I(I+1) - 3p^2). \tag{25}$$

The values of k are given in Table 1 and $p=-M$ for a coherence order $M=2I$, $p=M$ for $M<2I$. The centre of gravity is generally defined as in Eq. (24)

$$\delta_2 = \frac{\int \nu_2 I(\nu_2) d\nu_2}{\int I(\nu_2) d\nu_2}. \tag{26}$$

When the lineshapes are clearly resolved in the 2D spectra, each site can be projected separately by taking the sum of relevant slices. The quadrupolar parameters can be either simulated or extracted by moment analysis [49] or direct calculation [50]. When distributions of quadrupolar interactions and chemical shifts exist, lineshapes cannot always be properly defined. This situation is encountered many times in substances like glasses, ceramics and mesoporous materials. In such cases, the extraction of the NQCC and η is cumbersome, but it is still possible to obtain the value of the second order quadrupole effect $\left(\mathrm{SOQE} \equiv \chi\sqrt{1+\eta^2/3}\right)$. The values of δ_{cs} and SOQE can be extracted by inverting Eqs. (23) and (24), resulting in Eq. (27) and Eq. (28),

$$\delta_{cs} = -\frac{(1+k)A(I,-1)}{pA(I,-1)+A(I,p)}\delta_1 + \frac{A(I,p)+kA(I,-1)}{pA(I,-1)+A(I,p)}\delta_2 \tag{27}$$

$$\delta_Q = \frac{1+k}{pA(I,-1)+A(I,p)}\delta_1 + \frac{p-k}{pA(I,-1)+A(I,p)}\delta_2, \tag{28}$$

with

$$SOQE = \frac{\omega_0/2\pi}{10^3}\sqrt{\delta_Q}\times\frac{4I(2I-1)}{3}. \tag{29}$$

An alternative approach for the extraction of the NMR parameters is by performing MQMAS at two different magnetic fields, and comparing only the values of δ_1 along the F_1 frequency axis.

Experimental Aspects

Optimisation of MQMAS pulse sequence starts with a proper calibration of the rf power, so that selective pulses (selective 90° and 180°) could be employed accurately for z-filter or shifted echo experiments. This is done with a reference sample, usually a solution, where ν_Q is averaged to zero.

The MQMAS excitation and conversion pulses are then adjusted by monitoring the MQ echo intensity in a one-dimensional (1D) experiment, employing a very small delay, between the excitation and conversion pulses, of say, 5 μs. Initially, the excitation pulse is fixed to some value and the duration of the conversion pulse is varied. Then, the length of the excitation pulse is varied for the optimised conversion pulse length. The optimisation protocol can be usually accomplished on a sample of Na_2SO_4 or Na_2ClO_3 for spin-$\frac{3}{2}$ systems and aluminium acetyl acetonate for spin-$\frac{5}{2}$ systems, but it may be noted here that the optimum values are ν_Q dependent, Fig. 8.

Starting values for the pulses can be obtained from numerical simulations as illustrated in Fig. 8. Figure 8a shows the maximum 3QC intensity that could be obtained for various quadrupolar strengths as a function of the pulse duration with the rf field fixed at 95 kHz. In b, the efficiency of the 3Q → 1Q conversion pulse is

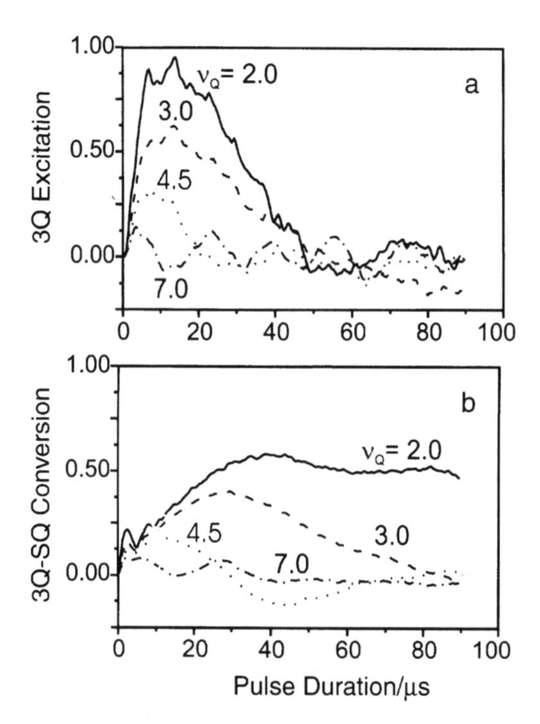

Fig. 8. (a) Plot of the 3Q intensity as a function of the duration of the pulse (excitation pulse in a MQMAS scheme) for various values of ν_Q and (b) plot of the SQ intensity as a function of the duration of the pulse (conversion pulse in a MQMAS scheme) for $\nu_1 = 95$ kHz, and MAS rate of 8 kHz. Three hundred powder orientations were considered for the simulation

indicated for each of the values of NQCC calculated in a. It may be emphasised here that the overall excitation efficiencies achieved by the optimised pulses are 62% and 33% for NQCC of 3.0 and 4.5 MHz with the amount of the 1QC eventually obtained being around 40% and 18%. The values are relative to the theoretical MAS single quantum signal and they already point to the main weakness of this method, which is the lack of an appreciable sensitivity.

In order to perform a successful and useful MQMAS experiment three steps need to be taken. They are (1) choosing a proper pulse scheme for pure phase spectroscopy, (2) selecting a suitable enhancement scheme to improve the sensitivity of the excitation and conversion pulses, and (3) optimisation of all the pulses. A pulse scheme should be chosen by taking into account several factors, some of which were already mentioned previously (relaxation times and spin quantum number) but also factors like the gyromagnetic ratio and the hardware capabilities. Some of these factors are crucial when choosing the right sensitivity enhancement scheme and will be elaborated on extensively in the next section.

Sensitivity Enhancement in MQMAS

In Fig. 8 it was shown that the excitation and conversion processes are inherently inefficient using hard pulses. The excitation efficiency of 3QC, for example, drops as $\frac{\nu_1^2}{\nu_Q^2}$. Some of the most important applications of MQMAS are envisaged in systems in which low-γ nuclei like ^{17}O, which is of low natural abundance also, can be probed routinely. It has also been established that a correlation of the highest possible MQC to 1QC achieves the best resolution, and if possible, MQC to NQC. However, excitation of 5QC, 7QC, 9QC in higher spin systems is even more difficult compared with the 3QC excitation. Figure 9 shows the amount of 3QC generated for a typical quadrupolar system for rf values of 50 kHz (a) and 100 kHz (b) as a function of the pulse duration for several MAS rates [51]. Although, high rf values (and high MAS for better lineshapes) are attractive with higher values of the NQCC, the required rf fields and MAS rates are out of reach in most of the spectrometers. Signal-to-noise can be gained by synchronous detection [52] or by acquiring multiple echoes during the free precession of the observable magnetisation (QCPMG) [53], but the gains achieved are not appreciable. The following sections demonstrate how a manipulation of the rf pulses can lead to significant signal gain which could of course be combined with the schemes mentioned above. The noticeable effect on the lineshapes is also discussed.

Enhancement Schemes for Spin-$\frac{3}{2}$

The first attempt to enhance 3Q excitation efficiency was by *Marinelli et al.* [54] making use of composite pulses. They showed that an application of two hard pulses with a 90° phase difference, and nutation angles of θ and 2θ for the first and second pulse respectively, resulted in an enhancement of the order of 30% in spin-$\frac{3}{2}$ systems. The flip angle θ needs to be optimised according to the rf power and quadrupolar strength. For sodium oxalate using $\frac{\nu_Q}{\nu_1} = 17$, for instance, the best 3QC was obtained with $\theta \approx \pi/2$. When large chemical shift anisotropies exist, such as

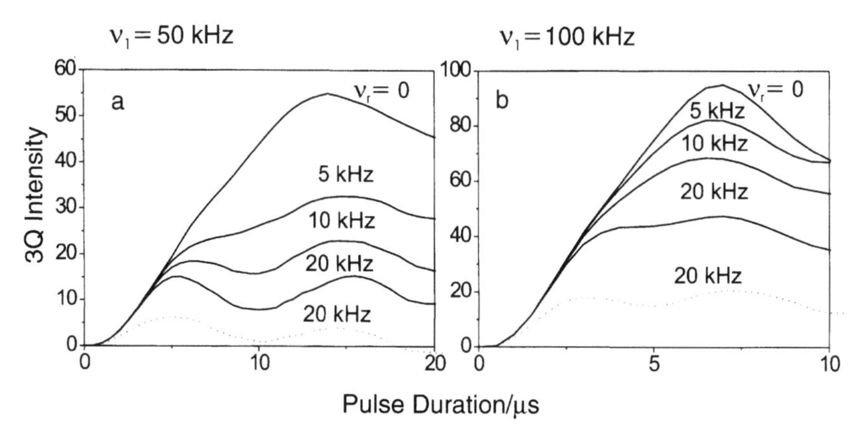

Fig. 9. Plot of the 3Q intensity (normalised with respect to single-quantum MAS signal) as a function of the duration of the pulse (excitation pulse in a MQMAS scheme) for (a) $\nu_1 = 50\,\text{kHz}$ and (b) $\nu_1 = 100\,\text{kHz}$ for various MAS rates, ν_r indicated on the top of each curve. The following quadrupolar parameters, $\nu_Q = 2.50\,\text{MHz}$ and $\eta = 0.70$, were assumed for the simulations except for the dotted curve for which $\nu_Q = 5.0\,\text{MHz}$ and $\eta = 0.70$

the case for a spin-7/2 ^{59}Co complex, a $x\bar{x}$ composite pulse proved to be more efficient, but flip angles can no longer be predicted in a simple manner.

A significant step towards enhancing the MQMAS signal was made with the introduction of the RIACT-II (rotation induced adiabatic coherence transfer) scheme [55] (Fig. 10a). Here the excitation pulse consists of a selective 90° pulse creating central transition coherences, immediately followed by a long hard CW pulse, 1/4 of a rotor period (τ_r) in length. For the conversion pulse again a pulse of duration $\tau_r/4$ is used. Another advantage of this scheme is its ability to perform equally well on a relatively large range of quadrupolar constants, thus making MQMAS more quantitative. However, the main drawback of the experiment is the distortion of the lineshapes along the MAS dimension [56]. It has been shown that moving the position of the rf offset has a substantial effect on the pulse efficiency. Hence, it can be used to selectively excite certain sites according to their quadrupolar strengths, while minimising the intensity of others [57], and thus may be used as a spectral editing technique.

The RIACT approach was soon followed by an experimental scheme using modulated rf fields for conversion of MQC to SQC. Two variants were introduced, namely, double frequency sweeps, DFS, Fig. 10(b) [58] and fast amplitude modulation, FAM, Fig. 10c [59, 60]. The resulting signal enhancement was approximately 3-fold relative to the original two-pulse (CW–CW) scheme (Fig. 5 or Fig. 7b, the split-t_1 version) depending on the quadrupolar strength. While DFS uses continuous modulation of the rf frequency, FAM utilises a discreet single amplitude modulation frequency. Their major advantage over the RIACT-II scheme (having approximately the same signal enhancement) is the improved lineshapes along the MAS dimension, thereby allowing more accurate determination of NQCC and η. Both DFS and FAM are ν_Q dependent.

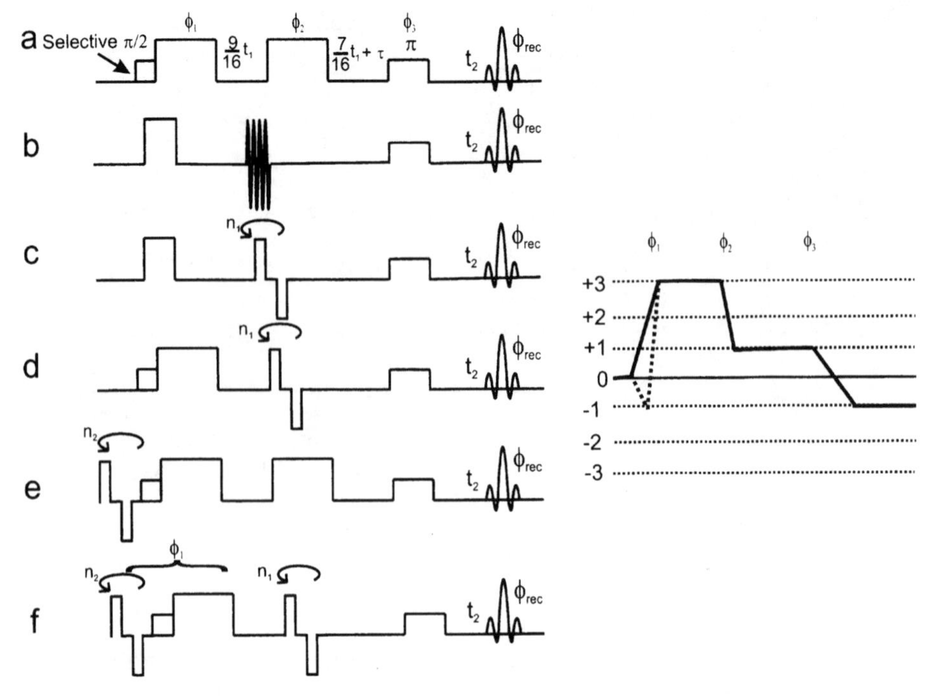

Fig. 10. Sensitivity enhanced MQMAS pulse schemes for spin-$\frac{3}{2}$ with split-t_1 whole-echo acquisition: (a) RIACT, (b) DFS, (c) FAM-I, (d) RIACT-FAM-I, (e) FAM-I-RIACT, and (f) FAM-I-RIACT-FAM-I sequences. The coherence pathways (dotted pathway for a, d, e, and f) selected for each of the schemes are the same as indicated in the figure; n_1 and n_2 are the number of times the FAM-I loop is repeated, with n_2 often equalling a full rotor period. Both $\frac{\pi}{2}$ and π pulses are soft pulses

The DFS pulses require an rf field strength of the order of 100–150 kHz. Diverging rf frequency sweeps are employed covering the whole ν_Q range. With no previous knowledge of the quadrupolar strengths, a sweep of over ≈ 2.5 MHz should be used, as this is normally the practical bandwidth of the probe. The sweep should start at a slightly off-resonance carrier frequency, starting with zero amplitude or half-Gaussian shape at the beginning of the pulse. Finally, one needs to optimise the sweep length. When the sweep covers the whole range of ν_Q values, an optimum length is approximately a quarter of a rotor period. These are starting values which can be further optimised [58, 61, 62].

FAM pulses, lately referred to as FAM-I, employ standard rf fields ($\nu_1 = 70$–110 kHz), and are designed from a repeating unit of four segments: a pulse with a positive phase, a delay, a pulse with a negative phase and a final delay $[\tau_x, \tau, \tau_{\bar{x}}, \tau]_n$, n is the number of repetitions). Optimisation is done by setting $\tau = 1$ µs, and $n = 4$. The values of n, τ are varied until maximum echo intensity is obtained in the 1D experiment. Variation of the basic FAM block can be made by adding several units with different n and τ numbers (extended FAM-I [63]), or by changing the lengths of the pulses with respect to the delays. The first option can be viewed as the discreet and simplified version of the DFS pulses. A major outcome of the introduction of these modulated rf schemes was the gaining of a better understanding of the MQC to SQC conversion process and of the characteristics of echo formation [60, 63]. This latter understanding resulted in the SL-FAM scheme, Fig. 10d (hereafter referred to as RIACT-FAM), obtained by combining RIACT excitation and extended FAM-I conversion, resulting in high sensitivity, relatively undistorted lineshapes and reduced dependence on NQCC.

With FAM-I it is possible to redistribute the population of the spin energy levels. This is called RAPT (rotor assisted population transfer) [64]. It has been shown that an enhancement by a factor of 1.5–2 is achieved in a MAS experiment on spin-$\frac{3}{2}$, when satellite redistribution is achieved before the excitation pulse, as shown in Fig. 11. The RAPT experiment comprised of the basic FAM-I scheme, $[\tau_x, \tau, \tau_{\bar{x}}, \tau]_n$, where the value of n is set to a large number such that the modulation lasts for a whole rotor period. The pulses are normally set to 1 µs and delays to the shortest time possible subjected to the hardware capabilities (e.g. transmitter phase stabilisation). However, shortening the delays is not very crucial for the performance of this scheme. The theoretical enhancement is $I + \frac{1}{2}$ if a complete satellite

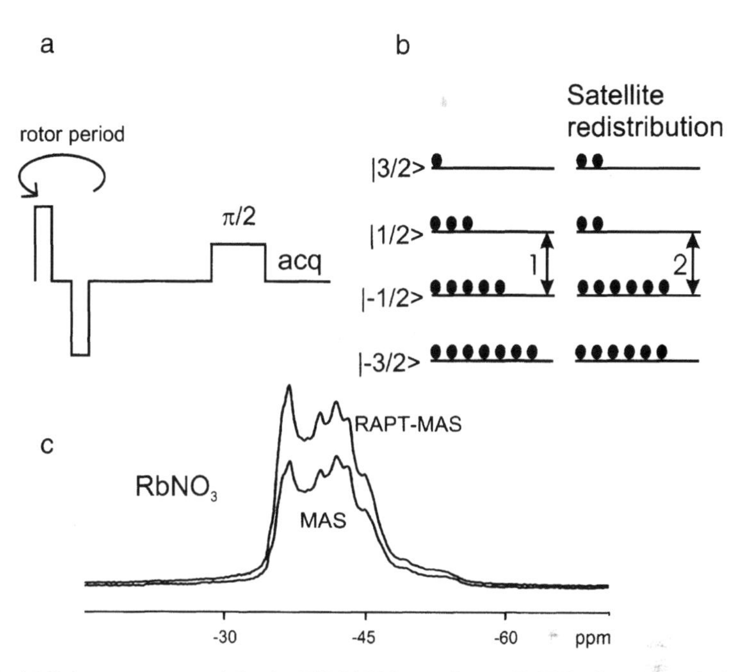

Fig. 11. (a) Pulse sequence used for the FAM-MAS experiment (RAPT scheme) employing FAM pulses lasting for a full rotor period followed by a soft $\frac{\pi}{2}$ pulse. (b) Population arrangement for a spin-$\frac{3}{2}$ system at thermal equilibrium (left) and after satellite redistribution with FAM pulses (right, one possible arrangement involving saturation of satellite transitions). (c) A single-pulse MAS spectrum and the corresponding FAM-MAS spectrum of a sample of $RbNO_3$. An enhancement by a factor of ≈ 1.8 was obtained with a spinning rate of 10.0 kHz and a FAM block given by [1 µs (pulse, x phase), 1 µs (delay), 1 µs (pulse, $\bar{x}$ phase), 1 µs (delay)], repeated 25 times lasting 100 µs. The external magnetic field was 4.1 T

saturation is achieved. The schematic in Fig. 11 corresponds to complete satellite saturation.

It is possible to combine RAPT with a MQMAS sequence that utilises SQC for the excitation of MQC and obtain an additional signal gain. The combination of RAPT and RIACT-II is shown in Fig. 10e [65]. A combination of RAPT with RIACT-FAM (FAM-RIACT-FAM) [66], Fig. 10f, gives the best performance so far for the MQMAS experiments in spin-$\frac{3}{2}$ systems, in terms of signal-to-noise and lineshapes.

Figure 12 shows the isotropic projection, anisotropic projection and a slice through one of the quadrupolar sites of a sample of $RbNO_3$ obtained by the use of six MQMAS schemes, namely, CW–CW, RIACT-II, CW-FAM-I, RIACT-FAM-I, FAM-RIACT-II and FAM-RIACT-FAM. Indeed, the last scheme provides the highest signal-to-noise ratio and the least distortions in the second-order quadrupolar

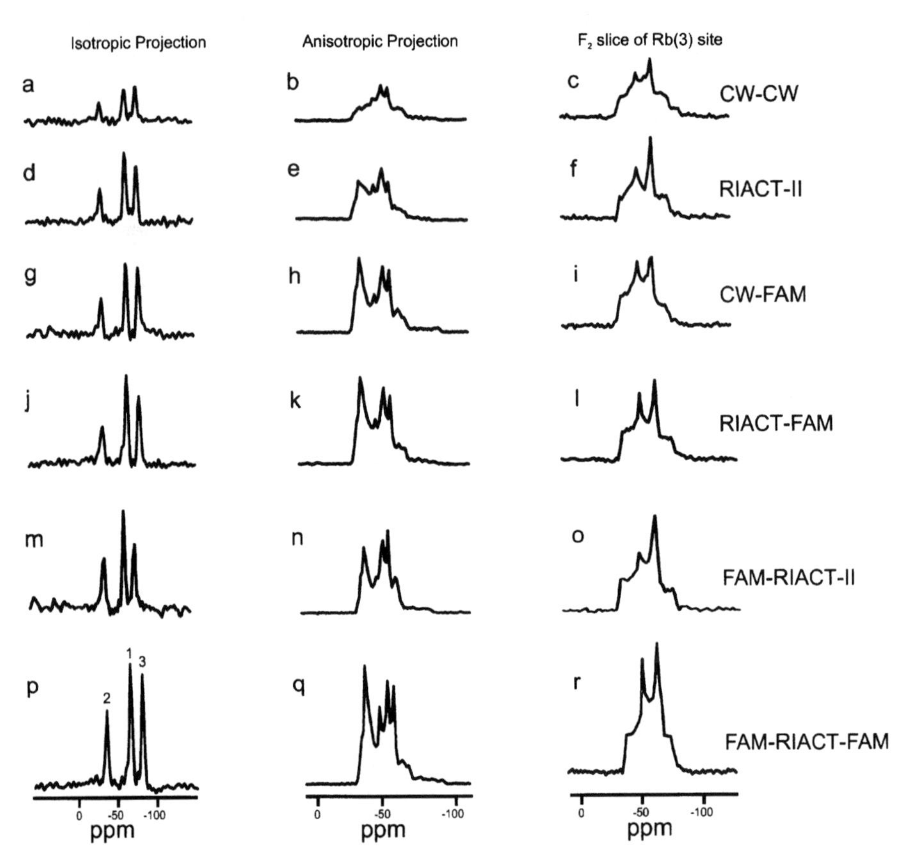

Fig. 12. Isotropic projection, anisotropic projection and an anisotropic slice across the Rb(3) site of $RbNO_3$ obtained from the pulse sequences CW–CW, RIACT, FAM-I, RIACT-FAM-I, FAM-I-RIACT and FAM-I-RIACT-FAM-I. For the experiments, an external magnetic field of 4.7 T, $\nu_r = 10.0$ kHz, hard pulses with a rf strength of 70.0 kHz, and soft pulses with a strength of 20.0 kHz were used. (Reproduced with permission from Ref. [66])

lineshapes. (No comparison was made with DFS scheme as a conversion pulse, however it is expected to perform at least equally as well as the FAM-I scheme).

FAM-I, DFS and RIACT use an adiabatic anti-crossing mechanism in order to efficiently convert the 3QC to the 1QC, th details of which are extensively dealt with in Ref. [60, 62].

A different approach to obtain sensitivity enhancement was suggested by *Vosegaard et al.* [67]. Two long pulses with a duration of 2–3 rotor periods were applied to the quadrupolar spins, employing low rf fields and high spinning rates. Rotary resonance conditions were found, in which an excitation minimum occurs when $2\omega_1 = n\omega_r$, ($n = 0, 1, \ldots$) and in between those minima points, enhanced signals were obtained. For the conversion pulse a maxima was found at $\omega_1 = n\omega_r$. This scheme called FASTER-MQMAS experiment achieves a 3-fold enhancement of the MQMAS spectra (like FAM-I, although at a lower rf power), but provides distorted lineshapes, as the excitation and conversion pulses select only a specific portion of the crystallites. A spectrum of $RbClO_4$ was obtained with a spinning rate of 30 kHz, an excitation pulse of duration $\tau = 75$ μs (2.25 rotor periods) and rf power $\nu_1 = 37$ kHz and a conversion pulse with $\tau = 65$ μs (1.95 rotor periods) and $\nu_1 = 30$ kHz. The selective echo π pulse was employed at $\nu_1 = 30$ kHz.

Enhancement Schemes for Spin-$\frac{5}{2}$

Spin-$\frac{5}{2}$ systems are more complex than spin-$\frac{3}{2}$ systems in the sense that additional energy levels are involved during the process of excitation and conversion. This fact complicates the rather simple mechanisms that explained the enhancement phenomena in spins-$\frac{3}{2}$. Thus, enhancement schemes for spin-$\frac{5}{2}$ based on the same design as spin-$\frac{3}{2}$ do not perform equally well.

Sensitivity enhancement of 3QMAS and 5QMAS spectra in spin-$\frac{5}{2}$ was obtained by employing FAM-I pulses for conversion of MQC to SQC [68]. *Iuga et al.* [61] have shown that a very accurate lineshape of a $\chi = 15.3$ MHz ^{27}Al site of the mineral andalusite could be obtained using DFS pulses from a 3QMAS spectrum. They also obtained some signal enhancement in the 5QMAS spectra of $9Al_2O_3 \cdot 2B_2O_3$ using DFS pulses, but with strong lineshape distortions [61]. Considerable shortening of the FAM-I pulse, up to a composite pulse scheme (FAM-II, Fig. 13a) [69] has been shown to enhance the 3QMAS signal of $AlPO_4$-5. This very short scheme makes FAM-II independent of the spinning rate. The same idea was exploited for the signal enhancement in 5QMAS experiments (Fig. 13b), where FAM-II was incorporated in the excitation pulse and either FAM-I or FAM-II used for conversion of 5QC to SQC [48]. A subtle point to be noted in Fig. 13a and b is the difference in the way split-t_1 scheme is executed for 3QMAS and 5QMAS in spin-$\frac{5}{2}$ systems. In general, for a correlation of the highest MQC with 1QC, a scheme like Fig. 13b needs to be chosen, while for a correlation of all the lower coherences with the SQC, a scheme like Fig. 13a is the desired one. FAM pulse scheme can not be efficiently incorporated into the original version of the z-filter scheme, but rather in a modified version in which the z-filter follows a split-t_1 procedure.

FAM-II is essentially composed of several pulses with alternating phases. The overall duration of a FAM-II pulse is short, up to ≈ 5 μs. A starting unit for $3Q \rightarrow 1Q$

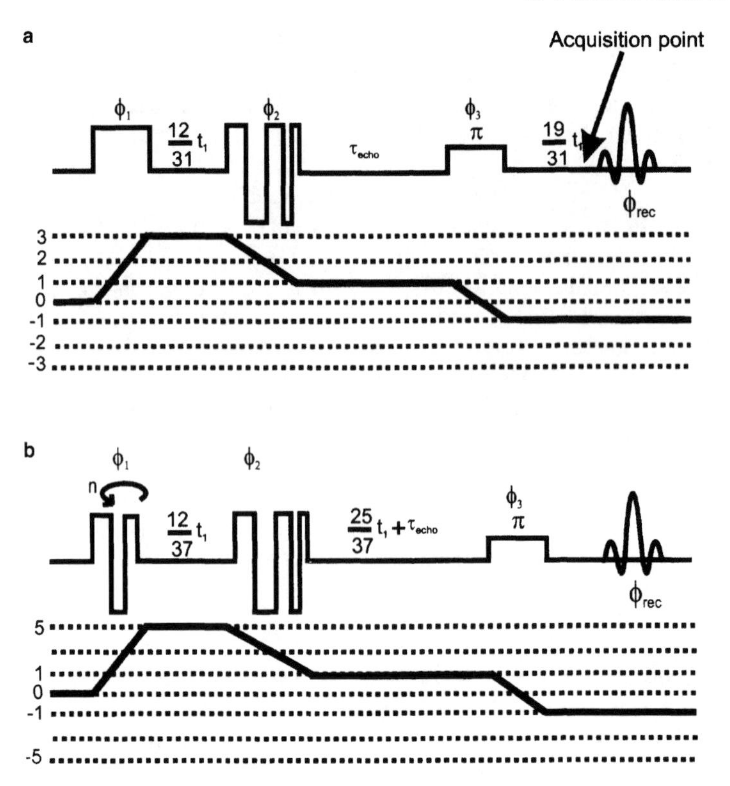

Fig. 13. (a) FAM-II split-t_1 whole-echo pulse sequence used for sensitivity enhancement in the 3QMAS of spin-$\frac{5}{2}$ systems. The acquisition starts after a delay of $\frac{19}{31}t_1$ (where the arrow points to). (b) CFF (CW-FAM-FAM) split-t_1 whole-echo pulse sequence used for the 5QMAS of spin-$\frac{5}{2}$ systems with FAM-II kind of approach employed for enhancing both excitation and conversion efficiency

conversion can be 2, $\bar{1}$, 0.5 μs. (A bar means a negative phase, numbers correspond to pulse durations). Addition or subtraction of segments, as well as length variations should be performed according to spectrometer capabilities, since transmitter phase glitches introduce pulse distortions.

In a 5QMAS experiment, a short $[x\bar{x}]_n$ modulation (FAM-II) is applied following a hard excitation pulse with optimised values of n and pulse durations. The hard excitation pulse creates 3QC from the equilibrium population and FAM-II converts the 3QC to 5QC. For conversion (5QC-SQC), a two-part FAM-I pulse or a two-part FAM-II pulse can be used. *Vosegaard et al.* suggested, for the first and second parts of a FAM-1 pulse, values of 0.4 and 0.85 μs and n of 6 and 4 [68]. A FAM-II pulse for conversion may be optimised from a series of pulses of the initial form $2, \overline{1.5}, 1, \overline{0.5}, 1.5, \bar{1}$.

Without loss of generality, it may be stated that FAM-I pulses are preferred for conversion of the highest MQC to 1QC, as this process is always adiabatic [60]. For all other MQC → 1QC, FAM-II pulses are preferred.

Although none of the enhancement schemes have been reported on higher spin systems (except for one example on cobalt, spin-$\frac{7}{2}$ [54]), it is reasonable to believe

Table 2. The phase table for 3Q- and 5QMAS split-t_1 whole-echo sequences

Experiment	Phase	Value/degrees
	ϕ_1	$(0, 30, 60, 90, 120, 180, 210, 240, 270, 300, 330)_8$
3QMAS	ϕ_2	$(0)_{96}$
	ϕ_3	$(0)_{12}, (45)_{12}, (90)_{12}, (135)_{12}, (180)_{12}, (225)_{12}, (270)_{12}, (315)_{12}$
	ϕ_{rec}	$\{(0, 270, 180, 90)_3, (90, 0, 270, 180)_3, (180, 90, 0, 270)_3, (270, 180, 90, 0)_3\}_2$
	ϕ_1	$(0, 18, 36, 54, 72, 90, 108, 126, 144, 162, 180, 198, 216, 234, 252, 270, 288, 306, 324, 342)_8$
5QMAS	ϕ_2	$(0)_{160}$
	ϕ_3	$(0)_{20}, (45)_{20}, (90)_{20}, (135)_{20}, (180)_{20}, (225)_{20}, (270)_{20}, (315)_{20}$
	ϕ_{rec}	$\{(0, 270, 180, 90)_5, (90, 0, 270, 180)_5, (180, 90, 0, 270)_5, (270, 180, 90, 0)_5\}_2$

that a gain in signal-to-noise could be obtained by using one or several of the above mentioned schemes.

We conclude this section by giving the explicit phase cycle values, Table 2, for split-t_1 shifted-echo 3QMAS (Fig. 7b, 10, 13a) and 5QMAS (Fig. 13b, 13c) sequences.

Heteronuclear Experiments Involving MQMAS

With MQMAS as a high-resolution solid state NMR tool, one may design heteronuclear experiments which lead to connectivity information, distance measurements and additional spectral editing techniques. Spin-locking of half-integer quadrupolar spins is hard to achieve. However, when the *Hartmann-Hahn* condition, $\omega_{S,nut} = \omega_{I,rf} \pm n\omega_r$ (n being an integer), is met, instances can be found in which cross-polarisation (CP) becomes possible to some extent [6, 7]. Several schemes have been proposed that combine CP with MQMAS. Initially it was suggested to cross-polarise ^{19}F to ^{27}Al [70] or ^{1}H to ^{27}Al [71] employing a very low rf field on the ^{27}Al channel, $\nu_{1S} = 5$ kHz, and to detect the z-filtered 3QMAS spectra of the ^{27}Al nucleus. These CP-MQMAS experiments gave very poor signal-to-noise ratio. While a normal 3QMAS experiment took appromimately 30 minutes to collect, a ^{27}Al{^{1}H}CP version had a poor signal intensity even after 16 hrs. Nevertheless, connectivity information could be obtained, thus providing spectral editing possibilities.

The efficiency of these experiments was improved by two different approaches. The first one uses again single-quantum cross-polarisation, combines the CP step with an inverse split-t_1 MQMAS experiment, which could incorporate FAM pulses to gain even more signal-to-noise [72]. The second one employs cross-polarisation directly to the triple-quantum transitions [73]. It was also demonstrated that any multiple-quantum transition can in principle be cross-polarised under suitable conditions [74, 75].

The cross-polarisation process was utilised to perform a MQMAS/HETCOR experiment [76], in which ^{23}Na polarisation was transferred to a ^{31}P nucleus after

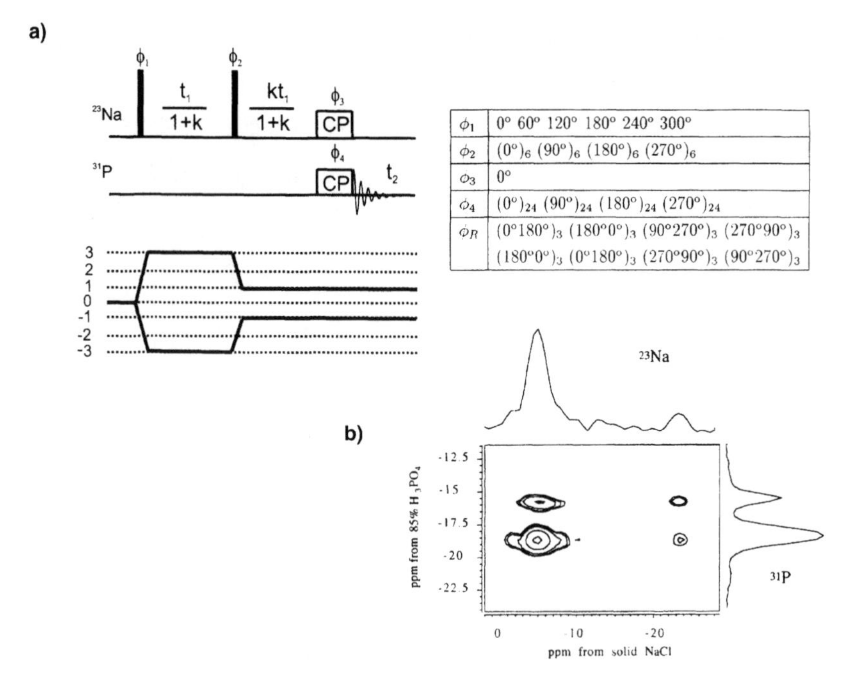

ϕ_1	0° 60° 120° 180° 240° 300°
ϕ_2	$(0°)_6$ $(90°)_6$ $(180°)_6$ $(270°)_6$
ϕ_3	0°
ϕ_4	$(0°)_{24}$ $(90°)_{24}$ $(180°)_{24}$ $(270°)_{24}$
ϕ_R	$(0°180°)_3$ $(180°0°)_3$ $(90°270°)_3$ $(270°90°)_3$ $(180°0°)_3$ $(0°180°)_3$ $(270°90°)_3$ $(90°270°)_3$

Fig. 14. (a) Pulse sequence, coherence transfer pathways and phase cycle table for MQMAS/HETCOR experiment. To obtain 2D pure-absorption line shapes both $+3$ and -3 pathways are retained. The 96 step phase cycle is such that both CYCLOPS [96] and spin-temperature alternation [97] are incorporated. The phase of ϕ_3 needs to be shifted by 90° to include STATES procedure. (b) ^{23}Na–^{31}P MQMAS-HETCOR spectrum of $Na_3P_2O_9$. The external magnetic field corresponded to a proton frequency of 500 MHz, the two pulses in (a) were 16 μs long corresponding to a 3π rotation on the central transition of ^{23}Na, CP contact time was 10 ms and $\nu_r = 5$ kHz. Thirty five complex t_1 increments were collected each consisting of 960 transients with a recycle delay of 3 s. (Reproduced with permission from Ref. [76])

performing a split-t_1 MQMAS experiment with amplitude modulated data set. In Fig. 14 an example of the pulse sequence and the corresponding spectrum is given. The advantage in transferring ^{23}Na magnetisation to ^{31}P in the sample of $Na_3P_3O_9$ and not *vice versa* stems from the relatively short T_1 of the ^{23}Na (3 s) compared with that of ^{31}P (600 s).

The rotational echo double resonance (REDOR) experiment [77] probes the distance between two spin-$\frac{1}{2}$ nuclei. It is possible to probe the distance to a quadrupolar nucleus by performing REAPDOR [78]. However, in this experiment, the quadrupolar spin is not directly detected. By applying REDOR decay pulses during SQ evolution (MQ-t_2-REDOR) [79] or during MQ evolution (MQ-t_1-REDOR) [80] of a MQMAS experiment, distances can be probed between a quadrupolar nucleus and a spin-$\frac{1}{2}$ nucleus, while detecting the quadrupolar spin. Application of the REDOR pulses during 3QC evolution, for instance, enhances the dipolar interaction by a factor of 3, thus enhancing the sensitivity of the experiment. Figure 15 demonstrates the applicability of the experiment on a sample of $AlPO_4$-CHA. The pulse sequence for (a) MQ-t_2-REDOR and (b) MQ-t_1-REDOR are shown, with a fit of distances in (c).

More recent attempts have been made in investigating residual dipolar couplings between quadrupolar nuclei [81], studying the cross-terms between quadrupolar and other interactions [82], elucidating relative orientations among quadrupolar nuclei [83] all of which are ultimately expected to yield more structural information.

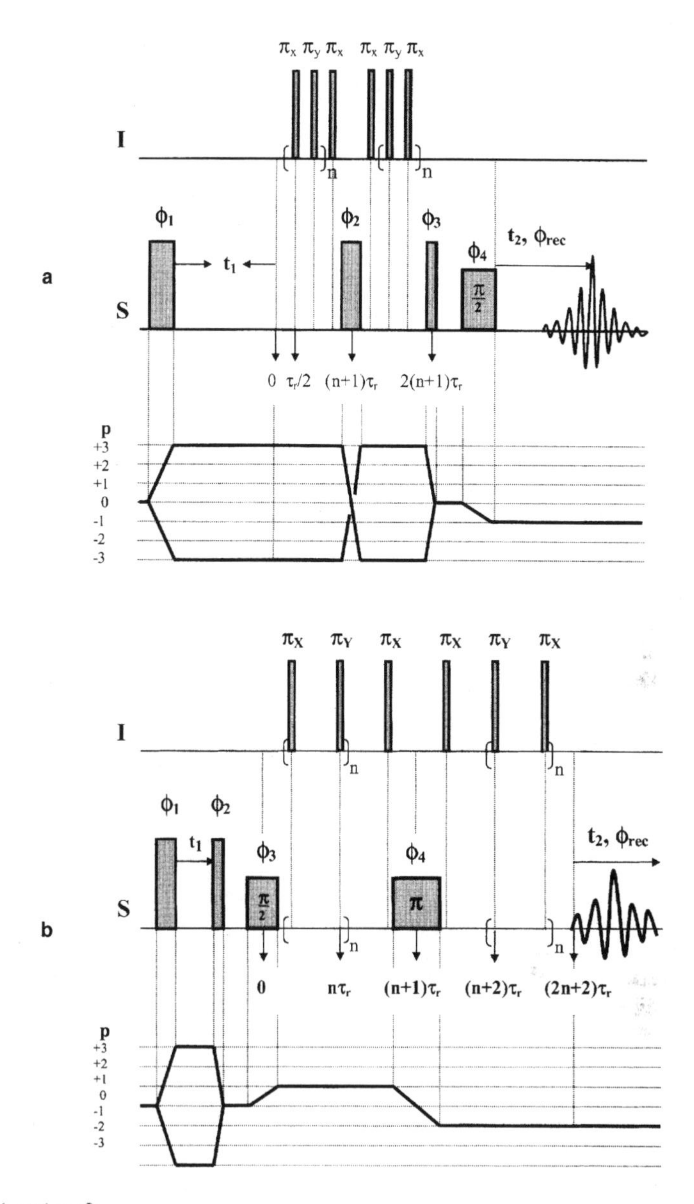

Fig. 15 *(continued)*

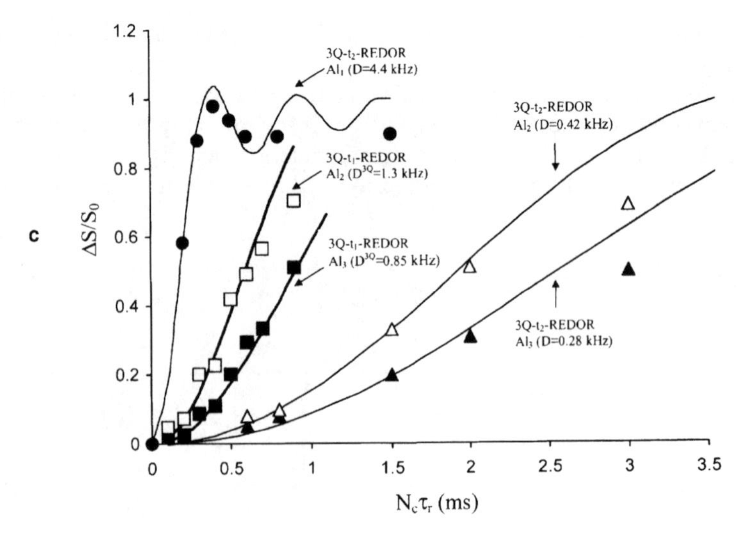

Fig. 15. (a) Pulse sequence used for 3Q-t_1-REDOR experiment. The phase cycle employed is as follows:

$\phi_1 = (0°)_{144}$,
$\phi_2 = (0°, 30°, 60°, \ldots, 330°)_{12}$,
$\phi_3 = [(0°)_{12}, (60°)_{12}, \ldots, (300°)_{12}]$,
$\phi_4 = (0°)_{72}, (180°)_{72}$,
$\phi_{rec} = [(0°, 180°)_6, (180°, 0°)_6]_3, [(180°, 0°)_6, (0°, 180°)_6]_3$.

A hypercomplex data set was collected by acquiring a complementary set of 96 fid's with $\phi_1 = 30°$. t_1 increments were rotor-synchronised.

(b) Pulse sequence used for 3Q-t_2-REDOR experiment. The phase cycle employed is as follows:

$\phi_1 = (0°)_{96}$,
$\phi_2 = (0°, 60°, 120°, 180°, 240°, 300°)_{16}$,
$\phi_3 = (0°)_{24}, (90°)_{24}, (180°)_{24}, (270°)_{24}$,
$\phi_4 = [(0°)_6, (90°)_6, (180°)_6, (270°)_6]_4$,
$\phi_{rec} = [(0°, 180°)_3, (180°, 0°)_3]_2, [(270°, 90°)_3, (90°, 270°)_3]_2, [(180°, 0°)_3, (0°, 180°)_3]_2, [(90°, 270°)_3, (270°, 0°)_3]_2$.

A hypercomplex data set was collected by acquiring a complementary set of 96 fid's with $\phi_1 = 30°$; t_1 increments were rotor-synchronised.

(c) Experimentally measured and simulated REDOR curves are shown for Al_1, Al_2 and Al_3 sites in $AlPO_4$-CHA obtained from 3Q-t_1-REDOR and 3Q-t_2-REDOR schemes. Analysis of the curves yielded the following distances: $r_{Al_3-F} = 4.1(\pm 0.1)$ and $r_{Al_2-F} = 4.7(\pm 0.1)$Å. (Reproduced with permission from Ref. [79], Coypright (1998), American Chemical Society and [80])

Common Applications

MQMAS has made study of various quadrupolar systems feasible in a routine way. Here, we outline a few representative examples of such investigations along with the type of information that may be extracted from MQMAS, possibly in conjunction with other experiments as well. The applications and observations are selected in a random order. The range of such studies is expected to increase with the advent of higher magnetic fields and by the use of sensitivity enhanced MQMAS schemes.

High Resolution ^{17}O NMR of Organic Solids

An important role for MQMAS is envisaged in the study of ^{17}O 3QMAS and 5QMAS experiments. This particular nucleus is expected to play a role in biomolecular structure determination with its high chemical shift dispersion. *Wu* [84] was the first to demonstrate the feasibility of such an experiment on some organic substances, [$^{17}O_2$]-D-alanine, potassium hydrogen [$^{17}O_4$] dibenzoate, [$^{17}O_4$]-D,L-glutamic acid-HCl and [2, 4-$^{17}O_2$] uracil. The 3QMAS spectrum of the glutamic acid is shown in Fig. 16. It is noted that substantial resolution can be achieved in these compounds with labelling of about 15–50% and moderate magnetic field of 11.75 T. Rotor synchronised t_1 increments were found to improve the signal-to-noise ratio. Besides providing information regarding isotropic chemical shift and quadrupolar parameters, ^{17}O MQ-MAS spectra are expected to shed light on several aspects of organic and biological systems, such as, H-bonding, side chain conformation, and base pairing in nucleic acids. One of the challenges in this field

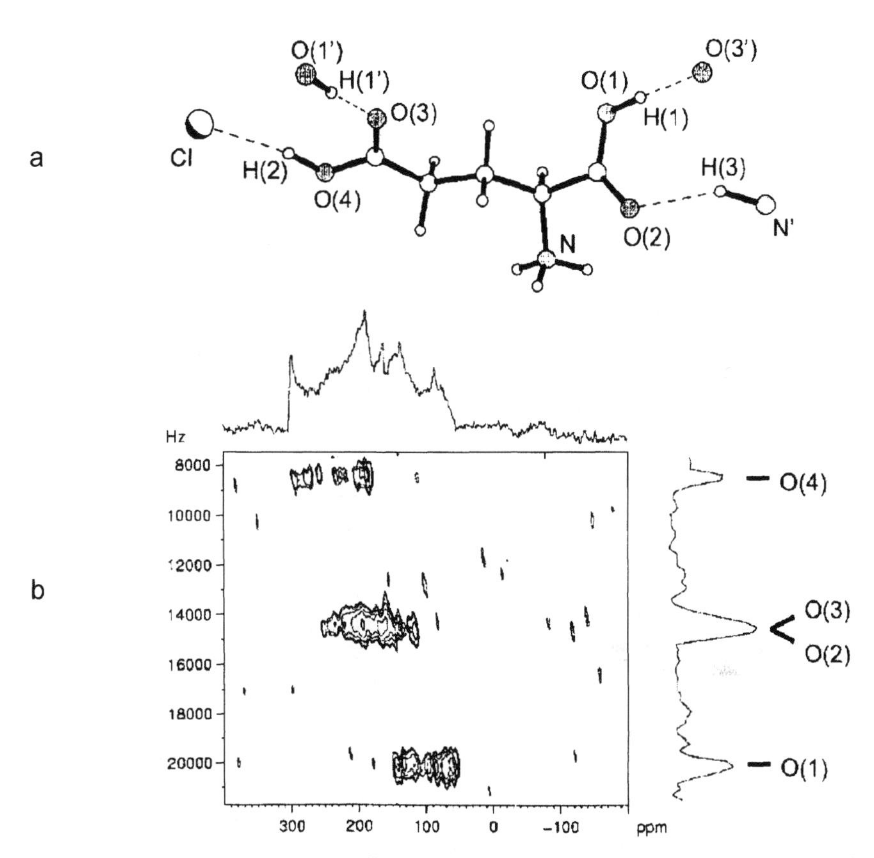

Fig. 16. (a) Molecular strucuture and ^{17}O positions marked of L-glutamic acid · HCL (b) ^{17}O 3QMAS spectrum of (a) with 28 rotor-synchronised t_1 increments. (Reproduced with permission from Ref. [84], Copyringht (2001), American Chemical Society)

lies in the synthesis of ^{17}O-enriched biological molecules together with the need to develop better sensitivity enhanced MQMAS schemes.

^{27}*Al MQMAS: Study of Porous Materials*

Zeolites play a big role in catalytic applications due to their stability and activity combined with a high selectivity. They are basically a unique class of porous solid aluminosilicates. ^{27}Al MAS NMR has been used in the past for a detailed study of aluminium coordination in zeolites [23, 85, 86]. However, the resulting spectra provide information that is normally restricted to aluminum coordination. Thus, resolving sites with similar environments is rarely achieved experimentally. ^{27}Al MQMAS, however, provides good resolution for the various ^{27}Al sites thereby allowing a near complete characterisation of such materials. When these materials are amorphous, details about the type of distribution become available. Properties like site specific dealumination [87] and positions and properties of acidic sites

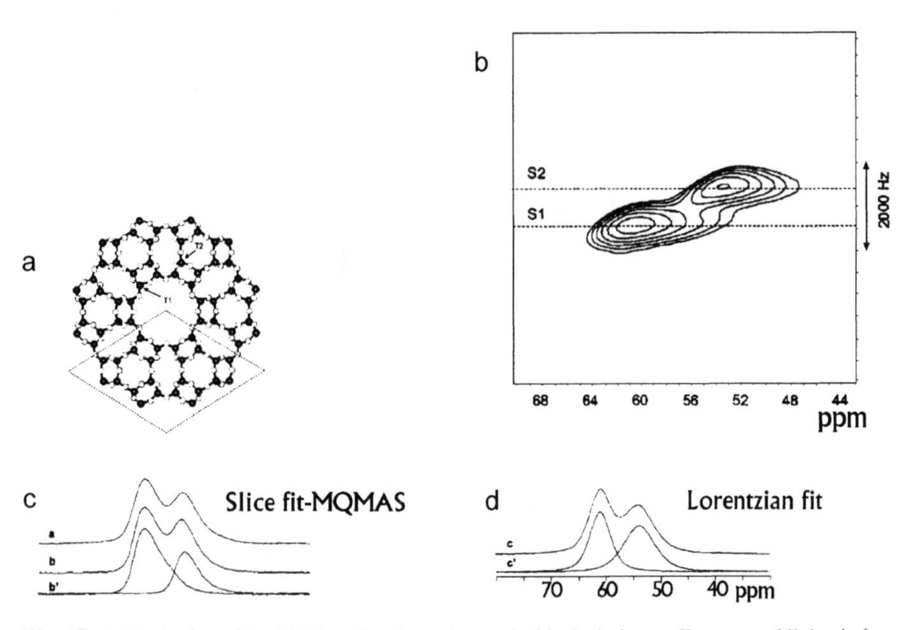

Fig. 17. (a) Projection of the MAZ zeolite along the *c*-axis, black circles are T atoms and light circles are the oxygen atoms. Tetrahedral framework aluminium sites are assigned to the Al atoms incorporated in the T1 and T2 crystallographic sites. (b) ^{27}Al 3QMAS spectrum of the MAZ zeolite sample with the horizontal axis corresponding to the F_2 dimension and the vertical axis corresponding to the F_1 dimension. The anisotropic slices corresponding to S1 and S2 are shown in (c) in the first row. Note the deviation of these lineshape from Lorentzian (d, first row) as assumed in the earlier studies. Also shown in (c) are the 1D ^{27}Al MAS spectrum of the MAZ zeolite (third row) and fitted spectrum (second row) using the slices S1 and S2 obtained from the 3QMAS spectrum. Also shown in (d) are the fitted spectrum (second row) using two Lorentzian shaped lines shown in the first row. A better fit may be found in (c) by virtue of the availability of true lineshapes due to resolution enhancement enabled by MQMAS spectrum. (Reproduced with permission from Ref. [90], Copyright (1999), American Chemical Society)

[88, 89] can be obtained unambiguously. *Wouters et al.* [90] demonstrated how MQMAS can give improved information about site populations. ^{27}Al MAS NMR studies have been used to arrive at the ratio of $\frac{Al^{T1}}{Al^{T2}}$ (corresponding to the two crystallographic sites, T_1 and T_2, Fig. 17a) in the mazzite zeolite. Slices along the MAS dimension of a 3QMAS spectrum (Fig. 17b) were used to deconvolute the 1D MAS spectrum (Fig. 17c), instead of the traditional Lorentzian/Gaussian lineshapes (Fig. 17d). The site population was shown to be completely different than previous reports (1.75 instead of 0.86) with a much better fit. It may be noted that, in general, site population estimates can be misleading in NMR experiments since the lineshapes and the positions of the lines are field dependent.

^{11}B MQMAS and MAS of Glasses

^{11}B MAS and MQMAS have been applied to a study of $x\mathrm{Na_2S} + (1-x)\mathrm{B_2S_3}$ glasses and polycrystals [91]. With the enhanced resolution from the MQMAS spectra, it

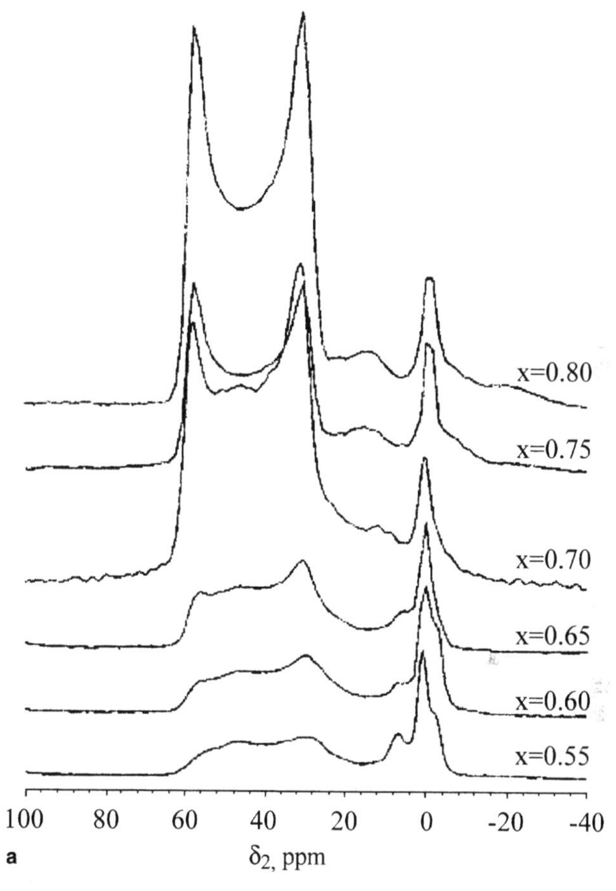

Fig. 18. (*continued*)

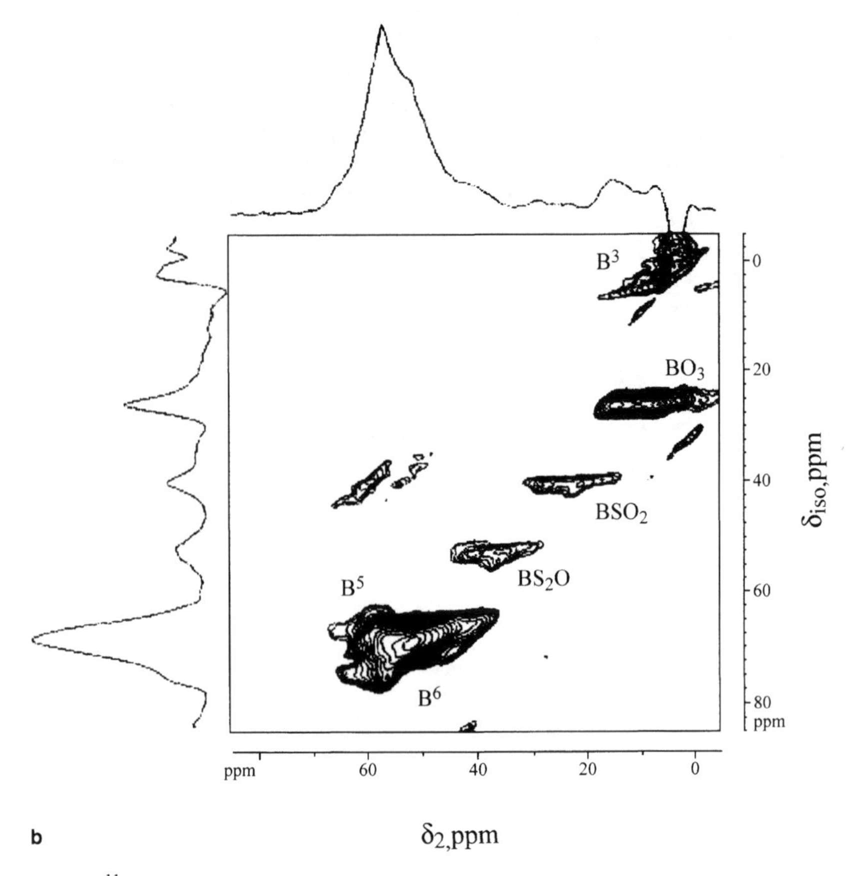

Fig. 18. (a) ^{11}B MAS NMR spectra at a magnetic field of 5.9 T of $x\mathrm{Na_2S} + (1-x)\mathrm{B_2S_3}$ glasses in the high alkali glass forming range $0.55 \leq x \leq 0.80$. (b) ^{11}B 3QMAS NMR spectrum taken at 9.4 T for the glass formed at $x = 0.6$. (Reproduced with permission from Ref. [91], Copyright (1998), American Chemical Society)

was possible to simulate corresponding MAS spectra and deduce isotropic chemical shift, quadrupolar parameters and relative concentrations of the various boron sites. Based on the known structure of boron trisulphide ($x = 0$), sodium methathioborate ($x = 0.5$) and sodium orthothioborate ($x = 0.75$), a model was suggested for the structure of sodium dithioborate ($x = 0.33$). In glasses, several structural units of the polycrystalline sample were observed alongside other new species. The evolution of the glass with an increasing $\mathrm{Na_2B}$ was tracked down and oxygen impurities were observed by MQMAS and MAS spectra with varying x values. The [23]Na spectra were not so informative due to the high mobility of the $\mathrm{Na^+}$ ions, producing broad NMR spectra. Figure 18 shows the evolution of the MAS spectra of a glass formation with x between 0.55 and 0.8 (Fig. 18a) and a MQMAS spectrum of a glass formed at $x = 0.6$ (Fig. 18b). The evolution towards an orthothioborate ($x > 0.75$, large powder pattern at 40–60 ppm) is clearly observed. The oxygen impurities are detected in the MQMAS spectra in addition to other basic structural units.

Additional studies have been performed on aluminosilicates utilising ^{17}O and ^{27}Al MQMAS NMR [93, 92] providing information about bond angle distributions and coordinations.

Observation of Indirect Spin–Spin Couplings in Quadrupolar Nuclei

The indirect spin–spin coupling constant, J, which is a useful parameter providing insight into molecular structure and chemical bonding, is difficult if not impossible to measure in solid state NMR of quadrupolar systems. The splitting arising from J will be masked in the second-order quadrupolar broadening. The advent of MQMAS has made the observation of J possible as demonstrated by *Wu et al.* [94] in the ^{11}B $\left(I = \frac{3}{2}\right)$ 3QMAS spectrum of a solid borane–triphenylphosphite complex, (PHO_3P–BH_3). Two isotropic peaks representing a doublet pattern were found. The experiments were performed at two magnetic fields to confirm that the splitting was indeed from one-bond J coupling between ^{11}B and ^{31}P. The experimental spectrum and derived NMR parameters are shown in Fig. 19. The apparent splitting in the indirect dimension of MQMAS spectrum is 2.125 J thus leading to an enhanced resolution. This enhancement is generally higher for the highest order of MQC. For lower MQC in higher spin systems one gets a reduction in the J splitting (except for 7QC in spin-$\frac{9}{2}$) [94].

Co-ordination Environments of ^{25}Mg in Nucleic Acids

Magnesium, ^{25}Mg, is a spin-$\frac{5}{2}$ nucleus with a low NMR sensitivity. MQMAS has been shown to be effective for the resolution and assignment of Mg binding sites

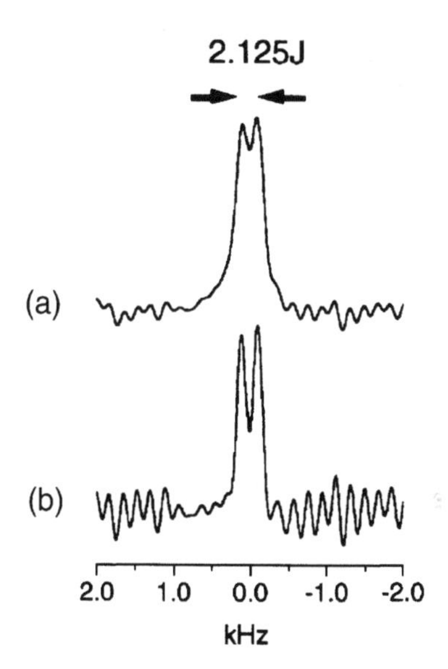

Fig. 19. ^{11}B 3QMAS spectra of $(PhO)_3P$–BH_3 at 7.4 T showing a J split doublet (a) with and (b) without resolution enhancement. The J splitting arises due to a $^{1}J(^{11}B, ^{31}P) = 85 \pm 5$ Hz as measured from the MQMAS spectrum. An analysis of the ^{11}B MQMAS spectra also yielded the following parameters: $\nu_Q = 1.22 \pm 0.02$ MHz, $\eta = 0.10 \pm 0.05$, and $\delta_{iso} = 3.0 \pm 0.1$ ppm. (Reproduced with permission from Ref. [94], Copyright (1994), American Chemical Society)

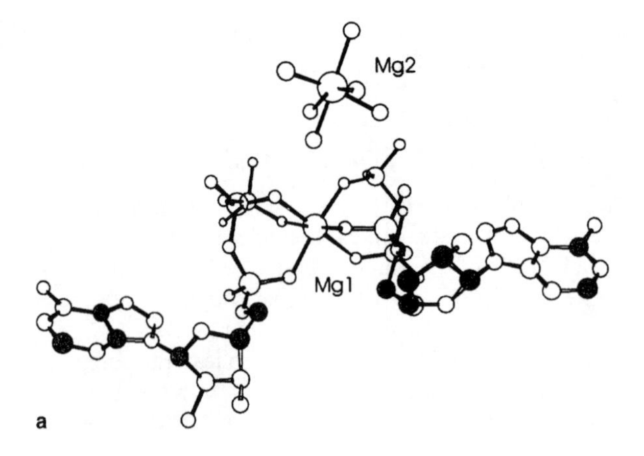

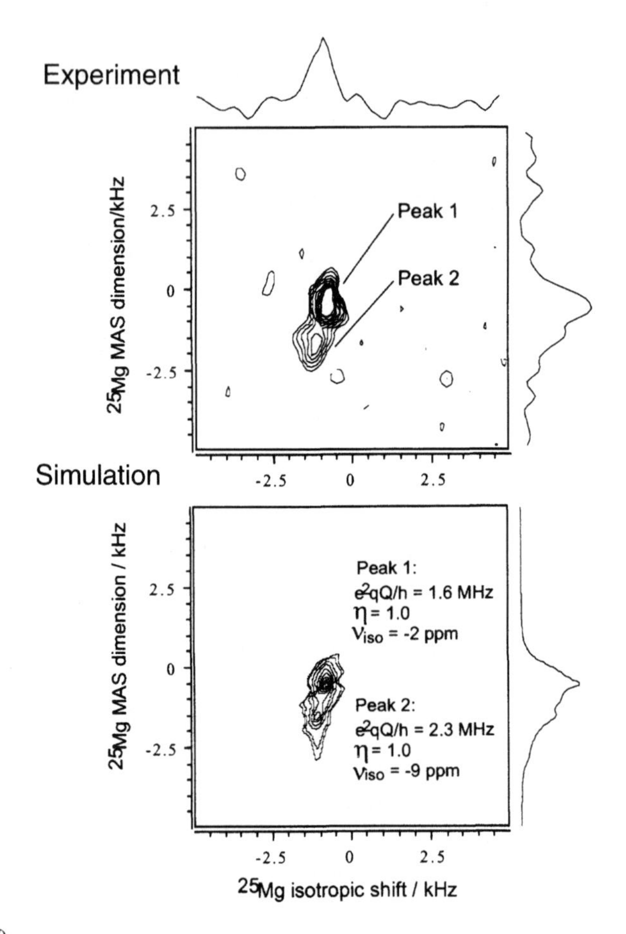

Fig. 20 (*continued*)

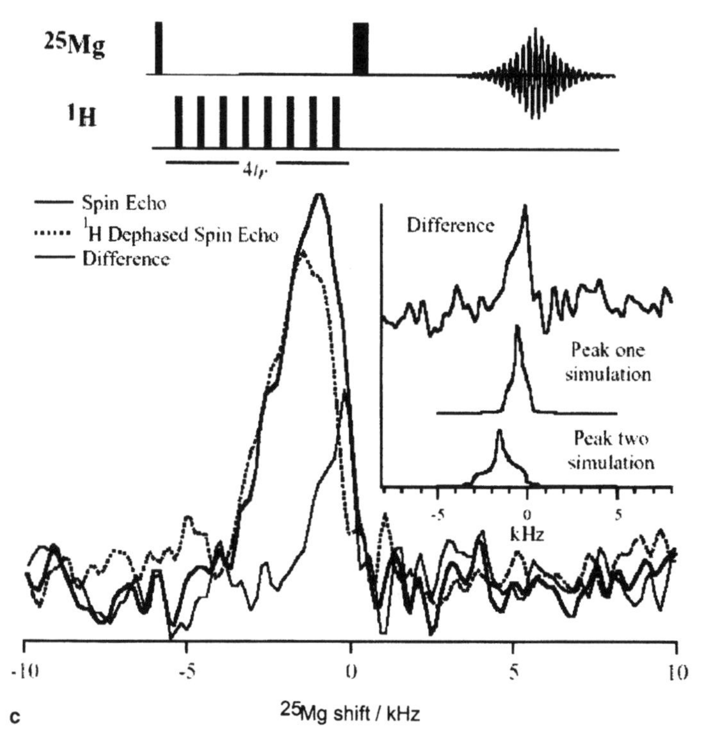

Fig. 20. (a) Illustration of the two magnesium coordination environments in MgATP/BPA. (b) Top spectrum is a ^{25}Mg 3QMAS spectrum of MgATP/BPA acquired with the following parameters, external magnetic field 11.7 T, $\nu_r = 10.0$ kHz, CW–CW pulse scheme with an excitation pulse of 7.5 µs and conversion pulse of 4.0 µs, number of transients of 24000 and number of t_1 increments of 16. Two magnesium peaks can be observed in the spectrum. The bottom spectrum is the best fit simulations of the experimental spectrum. The deduced parameters are indicated in the figure. (c) REDOR kind of pulse sequence and resulting difference curves that lead to the identification of which of the ^{25}Mg signal is being generated by each coordination environment shown in (a). The ^{25}Mg spin-echo data is obtained by the pulse sequence having $\frac{\pi}{2}$ and π pulses of duration 4.4 µs and 8.89 µs. An interpulse spacing of 400 µs had 8 ^{1}H π pulses inserted, centered at $\frac{1}{4}$ and $\frac{3}{4}$ of the rotor period. The inset graph compares the difference spectrum with the linseshapes simulated for peaks 1 and 2 using the parameters determined by MQMAS given in (b). (Reproduced with permission from Ref. [95], Copyright (2001), American Chemical Society)

within nucleic acids [95]. Figure 20a shows a ^{25}Mg 3QMAS spectrum of a ternary complex of magnesium(II), adenosine 5′-triphosphate, and bis(2-pyriyl)amine (BPA) ([Mg(H_2O)$_6$][HBPA]$_2$[Mg(MATP)$_2$] 12H_2O (MgATP/BPA), a schematic of which is shown in Fig. 20b. The two crystallographically inequivalent Mg sites were resolved in the spectrum, and the NMR parameters were derived. A combination of this data with a 1D proton REDOR dephasing experiment, Fig. 20c, enabled an assignment of the two Mg peaks. Peak 1 corresponds to Mg(2) co-ordinated by water molecules, and thus dominates the REDOR difference spectrum. The weakly coupled peak 2 corresponds to Mg(1) which is co-coordinated entirely by phosphate oxygen donors. The experiment mentioned above highlights the potential of

^{25}Mg NMR as a tool for elucidating magnesium coordination environments in biological complexes.

Conclusions

In this review an outline of the MQMAS experiment together with the basic theory for MAS and MQMAS of half-integer quadrupolar nuclei was given. Several experimental protocols were explored together with a detailed description of possible sensitivity enhancement schemes. For a spin-$\frac{3}{2}$, FAM-RIACT-FAM proved to be a very efficient enhancement scheme giving rise to relatively undistorted lineshapes. FASTER-MQMAS can be used when only low rf power is available (20–70 kHz), however, lineshapes are severely distorted. DFS pulses perform at least as well as FAM and give the user an additional free parameter (frequency modulation) to play with. DFS pulses can also be applied to low-γ nuclei with spin-$\frac{5}{2}$, since only low rf power levels are needed. When hardware is not suitable for these pulses, spin-$\frac{5}{2}$ MQMAS can be obtained with the aid of FAM-II pulses. The data analysis procedure was discussed in detail thus giving an accurate way to extract the relevant NMR parameters, *i.e.* NQCC and chemical shift. Heteronuclear experiments with high-resolution detection of the quadrupolar nucleus were explored and finally, some representative applications of the MQMAS experimental method were demonstrated.

Acknowledgements

The authors thank Prof. *S. Vega* for many discussions and suggestions on the manuscript. Discussions with Prof. *L. Frydman* are gratefully acknowledged. We thank Dr. *T. Bräuniger* and *M. Carravetta* for a critical reading of the manuscript. We thank Dr. *A. J. Vega* for providing us with the data which was used for the generation of Fig. 3.

References

[1] Abragam A (1961) The Principles of Nuclear Magnetism. Clarendon Press, Oxford
[2] Duer M (2002) Solid-State NMR Spectroscopy: Principles and Applications. Blackwell Science
[3] Pound RV (1950) Phys Rev **79**: 685
[4] Andrew ER, Bradbury A, Eades RG (1958) Nature **182**: 1659
[5] Lowe IJ (1959) Phys Rev Lett **2**: 285
[6] Vega AJ (1992) J Magn Reson **78**: 245
[7] Vega AJ (1992) Solid State NMR **1**: 17
[8] Samoson A, Lippmaa E, Pines A (1988) Mol Phys **65**: 1013
[9] Llor A, Virlet J (1988) Chem Phys Lett **152**: 248
[10] Chmelka BF, Müller KT, Pines A, Stebbins J, Wu Y, Zwanziger JW (1989) Nature **339**: 42
[11] Frydman L, Harwood JS (1995) J Am Chem Soc **117**: 5367
[12] Medek A, Harwood JS, Frydman L (1995) J Am Chem Soc **117**: 12779
[13] Gan Z (2000) J Am Chem Soc **122**: 3242
[14] Wu Y (1996) Encyclopedia of Nuclear Magnetic Resonance, Wiley, 1749
[15] Grandinetti PJ (1996) Encyclopedia of Nuclear Magnetic Resonance. Wiley, p 1768
[16] Gan Z (2001) J Chem Phys **114**: 10845
[17] Vega AJ (1996) Encyclopedia of Nuclear Magnetic Resonance. Wiley, p 3869

[18] Cohen MH, Reif F (1957) Solid State Phys 5. Academic Press Inc, p 321
[19] Rose ME (1957) Elementary Theory of Angular Momentum. Wiley, New York
[20] Zheng ZW, Gan Z, Sethi NK, Alderman DW, Grant DM (1991) J Magn Reson **95**: 509
[21] Man PP (1996) Encyclopedia of Nuclear Magnetic Resonance. Wiley
[22] Lippmaa E, Samoson A, Mägi M (1986) J Am Chem Soc **108**: 1730
[23] Kentgens APM (1997) Geoderma **80**: 271
[24] Vega S, Naor Y (1981) J Chem Phys **75**: 75
[25] Gerstein BC (1996) Encyclopedia of Nuclear Magnetic Resonance. Wiley, p 3360
[26] Ganapathy S, Schramm S, Oldfield E (1982) J Chem Phys **77**: 4360
[27] Jakobsen HJ, Skibsted J, Bildsøe H, Nielsen NC (1989) J Magn Reson **85**: 173
[28] Gan Z (2001) J Chem Phys **114**: 10845
[29] Dirken PJ, Kohn SC, Smith ME, van Eck ERH (1997) Chem Phys Lett **266**: 568
[30] Hwang SJ, Fernandez C, Amoureux JP, Han JCJW, Martin SW, Pruski M (1998) J Am Chem Soc **120**: 7337
[31] Fyfe CA, zu Altenschildesche HM, Skibsted J (1999) Inorg Chem **38**: 84
[32] Ferreira A, Lin Z, Rocha J, Morais CM, Lopes M, Fernandez C (2001) Inorg Chem **40**: 3330
[33] Lefebvre F, Amoureux JP, Fernandez C, Deronane EG (1987) J Chem Phys **86**: 6070
[34] Jerschow A, Logan JW, Pines A (2001) J Magn Reson **149**: 268
[35] Pike KJ, Malde RP, Ashbrook SE, McManus J, Wimperis S (2000) Solid State NMR **16**: 203
[36] Ernst R, Bodenhausen G, Wokaun A (1988) Principle of NMR in One and Two Dimensions. Oxford
[37] Bodenhausen G (1981) Prog Nucl Magn Reson Spectrosc **14**: 137
[38] Vega S, Pines A (1977) J Chem Phys **66**: 5624
[39] Wu G, Rovnyak D, Sun BQ, Griffin RG (1996) Chem Phys Lett **249**: 210
[40] Massiot D, Touzo B, Trumeau D, Coutures JP, Virlet J, Florian P, Grandinetti PJ (1996) Solid State NMR **6**: 73
[41] Schmidt-Rohr K, Spiess HW (1996) Multidimensional Solid State NMR and Polymers. Academic Press, pp 155–156
[42] Amoureux JP, Fernandez C, Steuernagel S (1996) J Magn Reson **A123**: 116
[43] Bax A, Mehlkopf AF, Smidt J (1979) J Magn Reson **35**: 373
[44] Brown SP, Heyes SJ, Wimperis S (1996) J Magn Reson **A119**: 280
[45] Brown SP, Wimperis W (1999) J Magn Reson **124**: 279
[46] Man PP (1998) Phys Rev B **58**: 2764
[47] Millot Y, Man PP (2002) Solid State NMR **21**: 21
[48] Goldbourt A, Vega S (2002) J Magn Reson **154**: 280
[49] Herreros B, Metz AW, Harbison GS (2000) Solid State NMR **16**: 141
[50] Engelhardt G, Koller H (1991) Magn Reson Chem **29**: 941
[51] Amoureux JP, Pruski M, Lang DP, Fernandez C (1998) J Magn Reson **131**: 17
[52] Massiot D (1996) J Magn Reson **A122**: 240
[53] Larsen FH, Skibsted J, Jakobsen HJ, Nielsen CN (2000) J Am Chem Soc **122**: 7080
[54] Marinelli L, Medek A, Frydman L (1998) J Magn Reson **132**: 88
[55] Wu G, Rovnyak D, Griffin RG (1996) J Am Chem Soc **118**: 9326
[56] Lim KH, Grey CP (1998) Solid State NMR **13**: 101
[57] Caldarelli S, Ziarelli F (2000) J Am Chem Soc **122**: 12015
[58] Kentgens APM, Verhagen R (1999) Chem Phys Lett **300**: 435
[59] Madhu PK, Goldbourt A, Frydman L, Vega S (1999) Chem Phys Lett **307**: 41
[60] Madhu PK, Goldbourt A, Frydman L, Vega S (2000) J Chem Phys **112**: 2377
[61] Iuga D, Schäfer H, Verhagen R, Kentgens APM (2000) J Magn Reson **147**: 192
[62] Schäfer H, Iuga D, Verhagen R, Kentgens APM (2001) J Chem Phys **114**: 3073
[63] Goldbourt A, Madhu PK, Kababya S, Vega S (2000) Solid State NMR **18**: 1
[64] Yao Z, Kwak H-T, Sakellariou D, Emsley L, Grandinetti PJ (2000) Chem Phys Lett **327**: 85

[65] Kwak H-T, Prasad S, Yao Z, Grandinetti PJ, Sachleben JR, Emsley L (2001) J Magn Reson **95**: 509
[66] Madhu PK, Levitt MH (2002) J Magn Reson **155**: 150
[67] Vosegaard T, Florian P, Massiot D, Grandinetti PJ (2001) J Chem Phys **114**: 4618
[68] Vosegaard T, Massiot D, Grandinetti PJ (2000) Chem Phys Lett **326**: 454
[69] Goldbourt A, Madhu PK, Vega S (2000) Chem Phys Lett **320**: 448
[70] Pruski M, Lang DP, Fernandez C, Amoureux JP (1997) Solid State NMR **7**: 327
[71] Fernandez C, Develoye L, Amoureux JP, Lang DP, Pruski M (1997) J Am Chem Soc **119**: 6858
[72] Ashbrook SE, Wimperis S (2001) Chem Phys Lett **340**: 500
[73] Ashbrook SE, Brown SP, Wimperis S (1998) Chem Phys Lett **288**: 509
[74] Ashbrook SE, Wimperis S (2000) J Magn Reson **147**: 238
[75] Rovnyak D, Baldus M, Griffin RG (2000) J Magn Reson **142**: 145
[76] Wang SH, De Paul SM, Bull LM (1997) J Magn Reson **125**: 364
[77] Gullion T, Schaefer J (1989) J Magn Reson **81**: 196
[78] Gullion T (1995) Chem Phys Lett **246**: 325
[79] Fernandez C, Lang DP, Amoureux JP, Pruski M (1998) J Am Chem Soc **120**: 2672
[80] Pruski M, Baily A, Lang DP, Amoureux JP, Fernandez C (1999) Chem Phys Lett **307**: 35
[81] Wi S, Frydman L (2000) J Chem Phys **112**: 3248
[82] McManus J, Kemp-Harper R, Wimperis S (1999) Chem Phys Lett **311**: 292
[83] Dowell NG, Ashbrook SE, McManus J, Wimperis S (2001) J Am Chem Soc **123**: 8135
[84] Wu G, Dong S (2001) J Am Chem Soc **123**: 9119
[85] Alemany ML (1993) Appl Magn Reson **4**: 179
[86] Smith ME (1993) Appl Magn Reson **4**: 1
[87] van Bokhoven JA, Koningsberger DC, Kunkeler P, van Bekkum H, Kentgens APM (2000) J Am Chem Soc **122**: 12842
[88] Kentgens APM, Iuga D, Kalwei M, Koller H (2001) J Am Chem Soc **123**: 2925
[89] Blumenfeld AL, Fripiat JJ (1997) J Phys Chem **101**: 6670
[90] Wouters BH, Chen T, Goossens AM, Martens JA, Grobet PJ (1999) J Phys Chem **B103**: 8093
[91] Hwang SJ, Fernandez C, Amoureux JP, Han JW, Cho J, Martin SW, Pruski M (1998) J Am Chem Soc **12**: 7337
[92] Baltisberger JH, Xu Z, Stebbins JF, Wang SH, Pines A (1996) J Am Chem Soc **118**: 7209
[93] Lee SK, Stebbins JF (2000) J Phys Chem **B104**: 4091
[94] Wu G, Kroeker S, Wasylishen RE, Griffin RG (1997) J Magn Reson **124**: 237
[95] Grant CV, Frydman V, Frydman L (2000) J Am Chem Soc **122**: 11743
[96] Hoult DI, Richards RE (1975) Proc Royal Soc (London) **A344**: 311
[97] Stejskal EO, Schaefer J (1975) J Magn Reson **18**: 560

Invited Review

Cross-Polarisation Applied to the Study of Liquid Crystalline Ordering

Krishna V. Ramanathan[1,*] and **Neeraj Sinha**[2]

[1] Sophisticated Instruments Facility, Indian Institute of Science, Bangalore-560012, India
[2] Department of Physics, Indian Institute of Science, Bangalore-560012, India

Received May 28, 2002; accepted June 19, 2002
Published online October 7, 2002

Summary. Cross polarisation is extensively used in solid state NMR for enhancing signals of nuclei with low gyromagnetic ratio. However, the use of the method for providing quantitative structural and dynamics information is limited. This arises due to the fact that the mechanism which is responsible for cross polarisation namely, the dipolar interaction, has a long range and is also anisotropic. In nematic liquid crystals these limitations are easily overcome since molecules orient in a magnetic field. The uniaxial ordering of the molecules essentially removes problems associated with the angular dependence of the interactions encountered in powdered solids. The molecular motion averages out intermolecular dipolar interaction, while retaining partially averaged intramolecular interaction. In this article the use of cross polarisation for obtaining heteronuclear dipolar couplings and hence the order parameters of liquid crystals is presented. Several modifications to the basic experiment were considered and their utility illustrated. A method for obtaining proton–proton dipolar couplings, by utilizing cross polarisation from the dipolar reservoir, is also presented.

Keywords. Cross-polarisation; Liquid crystals; Dipolar couplings; Ordering; Order parameter.

Introduction

Dipolar couplings have been one of the major sources of information for the study of liquid crystalline ordering [1–3]. While proton dipolar couplings have been used extensively in the case of small molecules with up to about ten spins oriented in liquid crystalline media, for larger molecules and for the liquid crystals themselves the method has been of very limited use. On the other hand, heteronuclear dipolar couplings pose a lesser problem in terms of line-width and resolution, and therefore offer an attractive means of studying liquid crystalline ordering. One of the first demonstrations of this approach has been made with the study of ^{2}H–^{13}C dipolar

* Corresponding author. E-mail: kvr@sif.iisc.ernet.in

couplings in *N*-(4-methoxy-benzylidene)-4-*n*-butylaniline (*MBBA*) deuterated at a specific site [4]. Similar studies have been performed on systems containing ^{19}F and using ^{19}F–^{13}C dipolar couplings [5, 6]. However, a more general and attractive approach would be to utilize ^{1}H–^{13}C dipolar couplings. The use of the 2D SLF procedure for this purpose was first demonstrated for the case of *MBBA* by *Hohener et al.* [4]. The ^{13}C chemical shifts and the corresponding ^{13}C–^{1}H dipolar splittings have been displayed along the F_2 and F_1 axes. However, in this method, the presence of the homonuclear dipolar couplings tend to make the lines broad along the F_1 axis. The problem of the elimination of the proton–proton dipolar couplings has been addressed by employing the variable angle sample spinning (VASS) technique [7]. Due to the off-magic angle spinning, the proton dipolar coupling is significantly reduced and further use of the multiple pulse decoupling during the t_1 period enables near elimination of proton dipolar couplings. The ^{13}C–^{1}H dipolar splittings observed along F_1 are to be suitably scaled to include the effect of VASS and the multiple pulse decoupling, to obtain the actual dipolar couplings. This method has been used for the study of a large number of liquid crystalline systems [8]. Modifications to this technique have also been suggested which enable short range [9] and long range [10, 11] dipolar couplings to be measured accurately.

Another approach that has been proposed and implemented is the use of a variant of the SLF 2D experiment that uses the transient oscillations observed during cross polarisation [12, 13] for estimating the dipolar couplings [14].

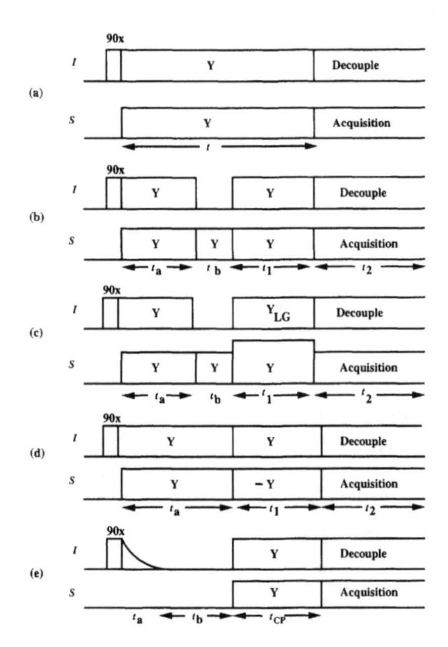

Fig. 1. Various cross-polarisation pulse schemes employed for obtaining dipolar couplings: (a) *Hartmann–Hahn* cross-polarisation, (b) cross-depolarisation, (c) cross-depolarisation with *Lee–Goldburg* decoupling, (d) polarisation inversion, (e) cross-polarisation from the dipolar bath

Modifications to the standard cross-polarisation pulse scheme such as the use of homonuclear decoupling during CP and the use of polarisation inversion have also been suggested [15, 16]. These are described in more detail in the following sections. A method of indirectly monitoring proton dipolar couplings using the carbon chemical shift dispersion has also been suggested which is also described. The various pulse schemes that have been utilized are shown in Fig. 1.

Dipolar Oscillations is Cross-Polarisation

Observation of dipolar oscillations during cross-polarisation (CP) has been made by *Muller et al.* in a single crystal of ferrocene [12]. In liquid crystals such as *MBBA* (Fig. 2), ^{13}C spectra containing sharp well-resolved lines can be obtained by the use of the CP sequence (Fig. 1a).

The intensity of the carbon lines show an oscillatory build-up when recorded as a function of the contact time t [14]. These oscillations are observed when the dipolar couplings of a carbon to its nearest neighbour protons are much stronger

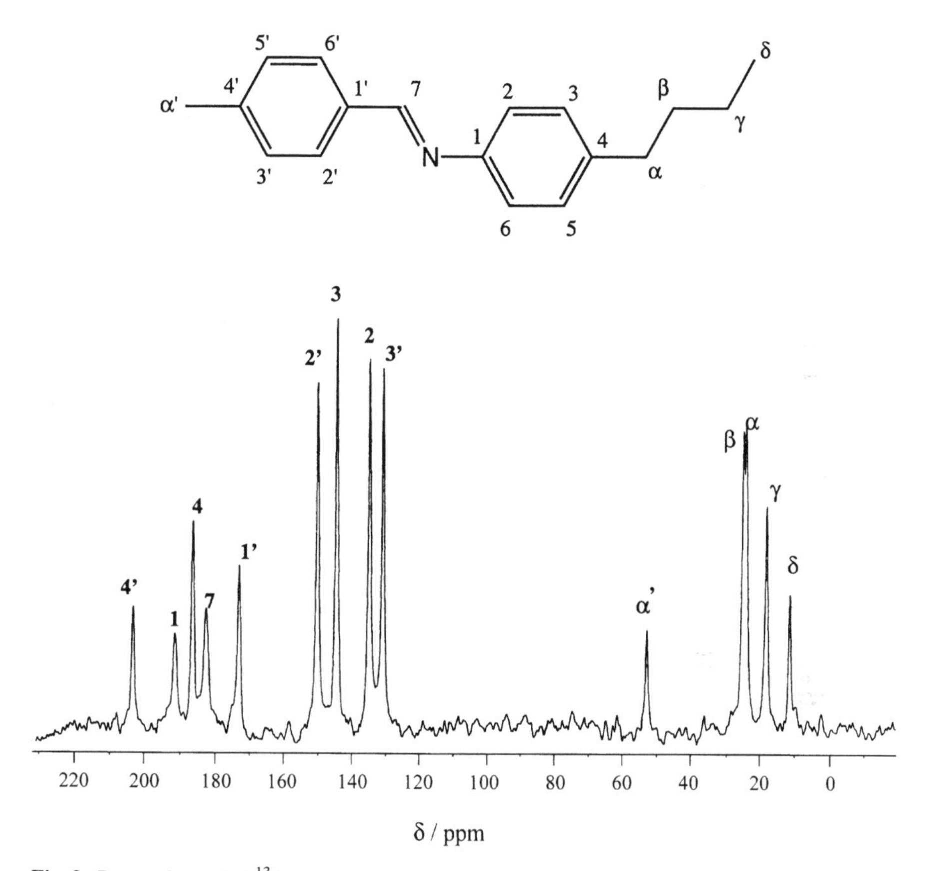

Fig. 2. Proton-decoupled ^{13}C spectrum of *MBBA* recorded on a Bruker DSX-300 NMR spectrometer at 75.47 MHz at room temperature

than the coupling of this spin system to the rest of the protons. The frequency f of the oscillations is given by

$$M_S = M_{S0}(1/2 - 1/2\cos(ft)). \tag{1}$$

The value of f is given in terms of D, the C–H dipolar coupling. It also depends on the number of equivalent protons coupled to a carbon. Thus for an IS spin system, such as a C–H group, $f=D$ and for the I_2S spin system such as the CH_2 group, $f=\sqrt{2}D$. For the methyl group (I_3S case), three values for f may be expected corresponding to D, $\sqrt{3}D$ and $2D$. Here the dipolar coupling D expressed in kHz is defined as

$$D = (-h\gamma_C\gamma_H/4\pi^2 r_{CH}^3)S_{CH} \tag{2}$$

where γ_H, γ_C are the gyromagnetic ratios of the protons and carbons respectively, r_{CH} is the internuclear vector and S_{CH} is the order parameter along the direction of this vector. The value of $(h\gamma_C\gamma_H/4\pi^2 r_{CH}^3)$ is 22.68 kHz for $r_{CH}=1.1$ Å. Equation (1) is further modified [12] to take into account spin diffusion between the protons directly coupled to the carbon under consideration and the rest of the protons in the system as

$$M_S = M_{S0}[1 - 1/2\exp(-t/T_{II}) - 1/2\exp(-3t/T_{II})\cos(ft)] \tag{3}$$

where T_{II} represents the time constant for the spin diffusion process.

Figure 2 shows the proton decoupled ^{13}C spectrum of *MBBA* at room temperature. Figure 3 presents the variation of the intensities of the C_7, C_α, C_β, and C_γ carbons as a function of the contact time. The oscillation frequency f and hence dipolar coupling D can be obtained by fitting Eq. (3) to the experimental data.

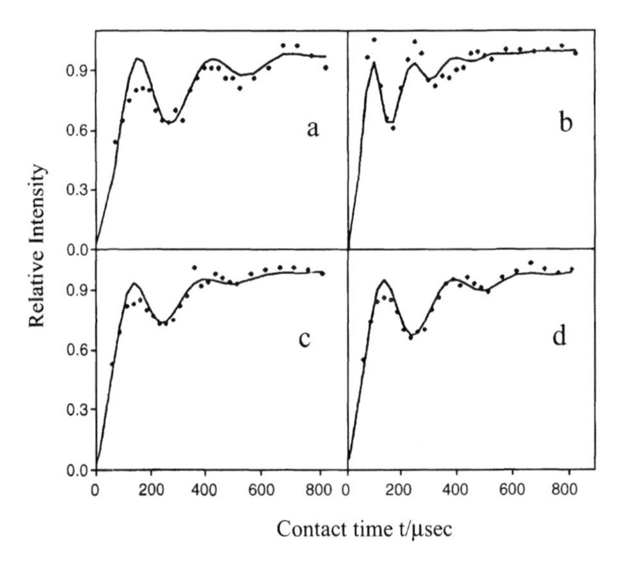

Fig. 3. Relative intensities of signals of (a) C_7, (b) C_α, (c) C_β, and (d) C_γ carbons of *MBBA* plotted against contact time. Points correspond to experimental values and the continuous line corresponds to the fit of the data to Eq. (3) with (a) $f=3.73$ kHz, $T_{II}=333$ μs, (b) $f=6.84$ kHz, $T_{II}=181$ μs, (c) $f=3.90$ kHz, $T_{II}=228$ μs, and (d) $f=4.0$ kHz, $T_{II}=273$ μs

Values of 3.73, 4.84, 2.76, and 2.84 kHz were obtained for D respectively for the C_7, C_α, C_β, and C_γ carbons of *MBBA*. The C_7 dipolar coupling enables the order parameter S of the aromatic core to be estimated with the assumption of uniaxial ordering from

$$D = -S(h\gamma_C\gamma_H/4\pi^2 r_{CH}^3)(3\cos^2\theta - 1)/2 \tag{4}$$

where θ is the angle between the director and the C–H bond. Using a value of 114° for the C–C–H bond angle and assuming a tilt of the director by 3.5° from the *para*-axis of the adjacent phenyl ring a value of $S = 0.52$ was obtained from these measurements, which agrees well with values reported by using other methods. The magnitude of the order parameter along the aliphatic chain averaged over several conformations can also be calculated from Eq. (2) using the obtained values of dipolar couplings. Values of 0.21, 0.12, and 0.13 obtained for the carbon sites α β, and γ are in agreement with the variation of order parameter generally observed in liquid crystalline systems.

A straightforward way of obtaining the oscillation frequencies is to use the 2D approach where the CP contact time (t) is incremented in regular steps between successive experiments. A two-dimensional data set is collected with proton decoupling during t_2. A two-dimensional *Fourier* transform then gives ^{13}C chemical shifts along the F_2 axis and the oscillation frequencies along F_1. A cross-depolarisation experiment (Fig. 1b), rather than a cross-polarisation experiment provides better results, the axial peaks which arise from the non-oscillatory terms of Eq. (3) is less intense in the former case (*vide infra*). The S spin magnetization during the t_1 period is given by

$$M_S = M_{S0}[1/2\exp(-t_1/T_{II}) + 1/2\exp(-3t_1/T_{II})\cos(ft_1)]. \tag{5}$$

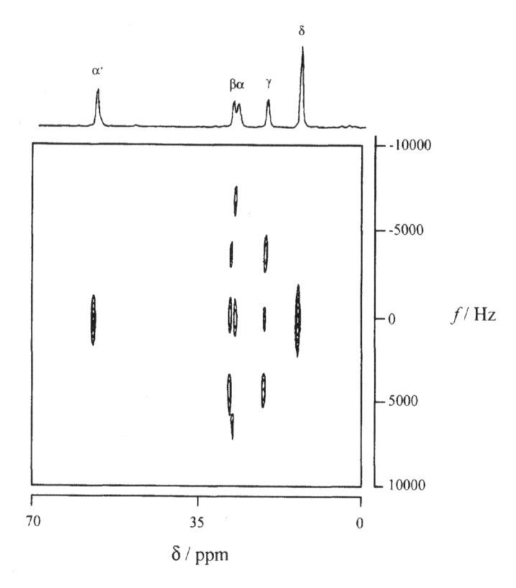

Fig. 4. Two-dimensional spectrum of the aliphatic carbons of *MBBA* obtained using the pulse sequence shown in Fig. 1b on a Bruker DSX-300 NMR spectrometer at 75.47 MHz at room temperature. The horizontal axis corresponds to ^{13}C chemical shifts and the vertical axis to the dipolar oscillation frequencies. The projection along the horizontal axis of the two-dimensional spectrum is also shown

The peaks along the F_1 axis provide the oscillation frequencies f from which the dipolar couplings can be estimated using the prescription provided earlier. Results of such an experiment carried out with $t_a = 1$ ms and $t_b = 200$ μs are shown in Fig. 4 for the aliphatic carbons of *MBBA*. There are cross peaks occurring at 6.8, 3.8 and 4.0 kHz corresponding to the C_α, C_β, C_γ carbons, which provide the corresponding C–H dipolar couplings as 4.8, 2.7, and 2.8 kHz.

Improved Schemes

Inclusion of homonuclear decoupling during CP

The dipolar oscillations during CP are highly damped due to the couplings among the protons. Consequently the peaks in the two-dimensional spectra are broad and in several instances hard to observe. *Wu et al.* [17] have shown that the use of *Lee–Goldburg* (*LG*) decoupling [18] during cross-polarisation results in the removal of homonulcear dipolar couplings leading to a reduction of the line width along the dipolar axis. The use of the method has been made for the case of the liquid crystal *N*-(4-ethoxybenzylidene)-4-*n*-butylaniline (*EBBA*) [15]. The pulse sequence used for the 2D experiment is shown in Fig. 1c. The t_1 period corresponds to transfer of polarisation from carbons to protons. During this period the proton offset is changed to satisfy the *Lee–Goldburg* condition and carbon power level is adjusted to satisfy the *Hartmann–Hahn* condition so that magnetization exchange takes place under homonuclear proton spin decoupling. The effect of *LG* decoupling resulting in the lengthening of the dipolar oscillation is shown for the case of the benzylidene carbon of *EBBA* in Fig. 5 where results of two experiments, *viz.*, one in which the proton r.f. is applied on-resonance corresponding to normal depolarisation experiment and the other in which *LG* decoupling is employed, are compared.

The homonuclear decoupling during the t_1 period results in the scaling of the heteronuclear couplings, the theoretical scaling factor being $\sin(\theta_m) = 0.82$, where θ_m is the magic angle. The two-dimensional spectrum of *EBBA* was obtained at

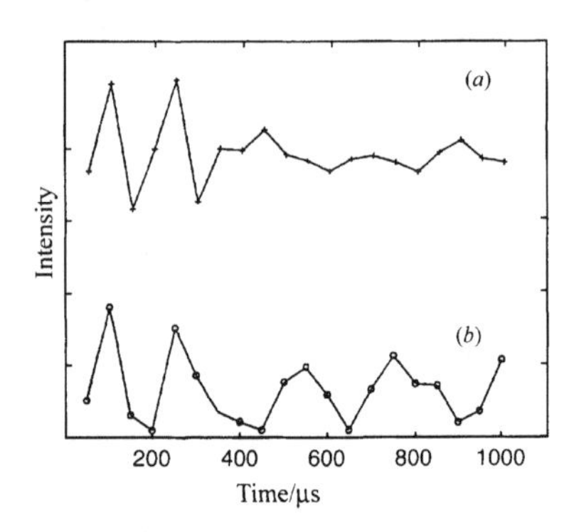

Fig. 5. Intensities of the benzylidene carbon of *EBBA*, as a function of t_1 obtained after *Fourier* transformation along t_2 of the two-dimensional data-set obtained using the pulse scheme shown in Fig. 1c; (a) with on-resonance proton r.f. during t_1 and (b) with *LG* decoupling during t_1

several temperatures using the method and C–H dipolar couplings at several sites were obtained. From the dipolar couplings of the benzylidene carbon, the order parameter of the aromatic core was obtained at these temperatures. Oscillation frequencies were obtained for the aromatic carbons also. From the values of order parameter of the core at different temperatures, the proton–carbon dipolar couplings of the aromatic core can be calculated, assuming a hexagonal geometry of the phenyl ring and fast flip motion about the para axis. In this case, the one bond C–H dipolar coupling $D^1_{\rm CH}$ is nearly the same as the coupling of this carbon to its ortho-proton $D^2_{\rm CH}$ due to the effect of ordering being different along different directions. In such cases, where the carbon is coupled to more than one proton, the evolution of magnetization is expected to proceed in a locked mode [13] and the oscillation frequency is given by $\sqrt{[(D^1_{\rm CH})^2 + (D^2_{\rm CH})^2]}$. The values thus calculated and the experimental oscillation frequencies are observed to show a good correlation [15].

Polarisation inversion

As pointed out earlier, Eq. (3) governs the frequencies and intensities of the peaks along the dipolar axis in the 2D experiments. The oscillatory cosine term gives rise to the cross peaks containing dipolar coupling information. There is also the non-oscillatory part which gives rise to the zero frequency peaks, which may overlap with cross peaks close to the center, causing difficulties in measuring small dipolar couplings. For attenuating the zero frequency peaks, use of polarisation inversion instead of cross polarisation has been utilized [16, 17]. The polarisation transfer process for a two spin I-S system can be modeled as a coherent process in mutually commuting zero-quantum and double-quantum manifolds [19]. Under the assumption of high r.f. fields, the dipolar coupling causes as oscillatory evolution of the density matrix in the zero quantum frame, while in the double quantum frame the density matrix remains constant. The former gives rise to the cross peaks and the latter to the axial peak. In polarisation inversion (Fig. 1d) there is an initial polarisation transfer from I spins to S spins during t_a. At the end of this period the two spin system can be taken to be completely polarised due to contact with the proton bath. Thus the initial density matrix $I_Z = I_Z^\Sigma + I_Z^\Delta$ tends to become $I_Z + S_Z = 2I_Z^\Sigma$ where $I_Z^\Sigma = 1/2(I_Z + S_Z)$ and $I_Z^\Delta = 1/2(I_Z - S_Z)$ are the density operators in the double and zero quantum frames respectively. Inversion of the r.f. field for the S spin corresponds at this point to a change of sign of the S spin Hamiltonian or equivalently the density matrix can be thought of as equal to $I_Z - S_Z = 2I_Z^\Delta$. As a result the initial evolution of magnetization during polarisation inversion takes place only in the zero-quantum sub-space. Therefore the dipolar cross-peaks are much more intense than the axial peaks enabling even small dipolar couplings to be observed. At larger contact times, the double-quantum evolution does come into play [16] and the equation governing this process is given by

$$M_S = M_{S0}[1 - \exp(-t_1/T_{II}) - \exp(-3t_1/T_{II})\cos(ft_1)]. \tag{6}$$

Figure 6 shows the contribution to the S spin intensity from the zero quantum and the double quantum evolution as well as the combined effect for both

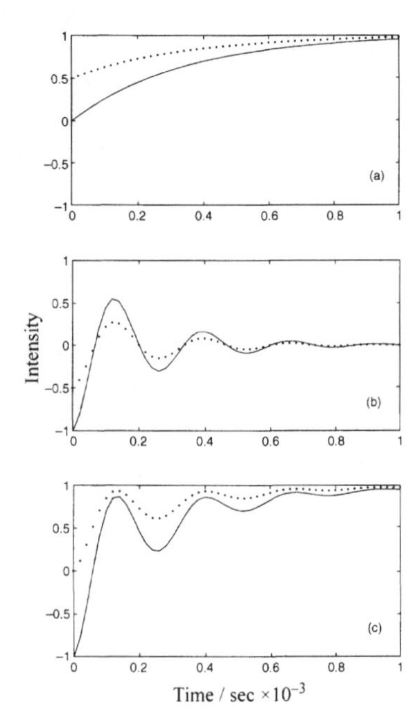

Fig. 6. S spin magnetization evolution due to (a) double quantum (b) zero quantum processes, and (c) combined effect of results shown in (a) and (b) during polarisation (····) and polarisation inversion (—). The plots correspond to values of f and T_{II} being 3.73 kHz and 333 μs

cross-polarisation and polarisation inversion experiments. It is observed that polarisation inversion leads to a doubling of the amplitude of the oscillatory part and also to a reduction in the initial value of the non-oscillatory component. In a 2D experiment, this would lead to a doubling of the cross-peak intensity and a significant reduction of the axial peak intensity.

A simulation of the relative intensities of the cross-peaks in comparison to the axial peaks for three experiments, namely (i) cross-polarisation, (ii) cross-depolarisation and (iii) polarisation inversion are shown in Fig. 7, using respectively Eqs. (3), (5), and (6). A dipolar coupling of 9.73 kHz and two different values of T_{II} equal to 330 μs and 1 ms were considered. It was observed that polarisation inversion provides the highest relative cross peak intensity for longer values of T_{II} considered, while cross-depolarisation provides higher cross-peak intensity in comparison to cross-polarisation for shorter T_{II} values.

Experimental demonstration of the usefulness of polarisation inversion has been carried out for the case of *MBBA* [16]. 2D spectra obtained using cross polarisation and polarisation inversion are displayed in Fig. 8. In the spectrum obtained using polarisation inversion shown in Fig. 8a, most of the carbon–proton dipolar couplings are observed to be resolved. For several carbons the cross-peaks are much more intense than the axial peaks such that they are not seen in the plot shown in Fig. 8a. Some typical cross-sections are shown in Fig. 9. It is interesting to note that not only short range dipolar couplings of carbons with attached protons

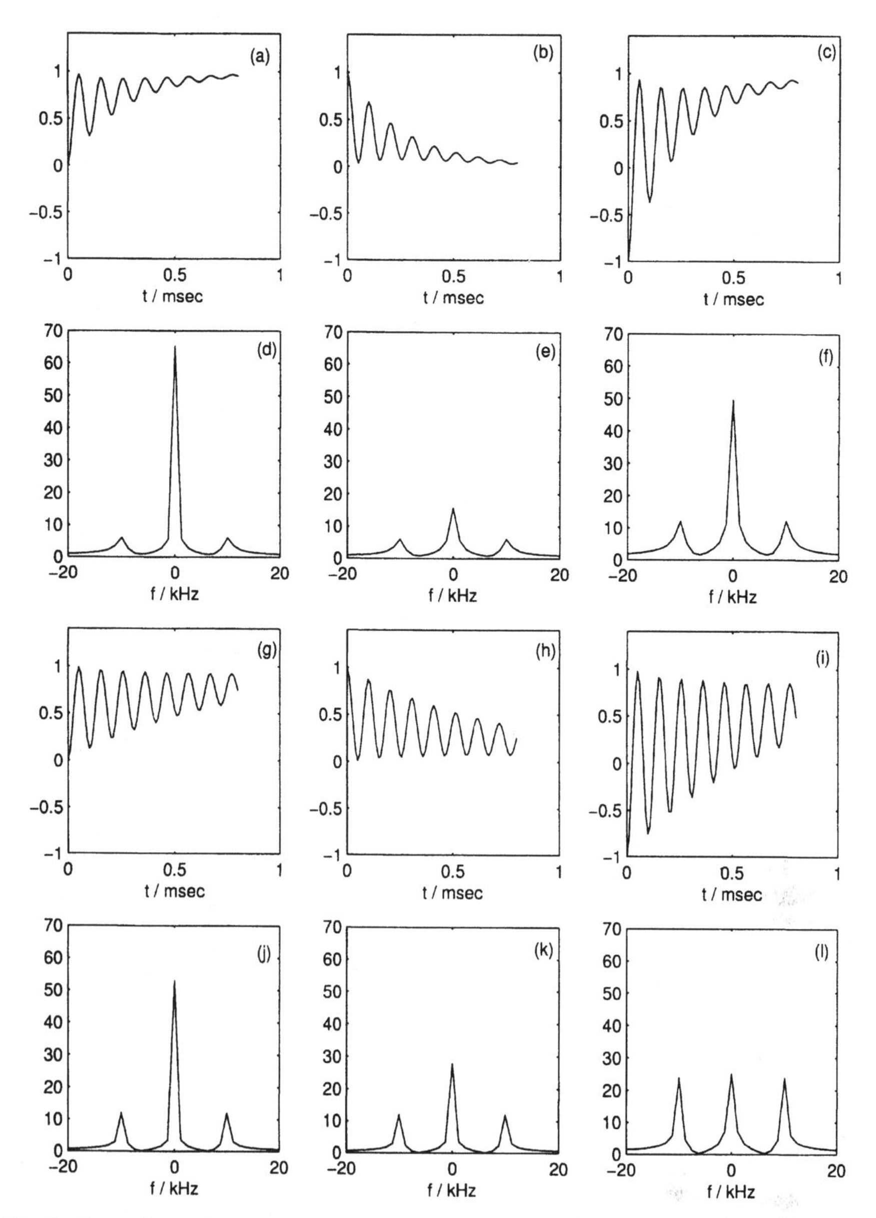

Fig. 7. Comparison of cross-peak intensity versus axial peak intensity for three different cross-polarisation experiments. The time domain signals have been obtained using Eqs. (3), (5), and (6). The corresponding *Fourier* transforms are shown below each. The first, second, and third columns correspond to polarisation, depolarisation, and polarisation inversion experiments. The dipolar coupling used is 9.73 kHz for all the cases. T_{II} is 330 μs for the top two rows and 1 ms for the two bottom rows

could be obtained, but also those of quaternary carbons coupled to remote protons are resolved. This then will lead to useful information on the order parameters of the system studied.

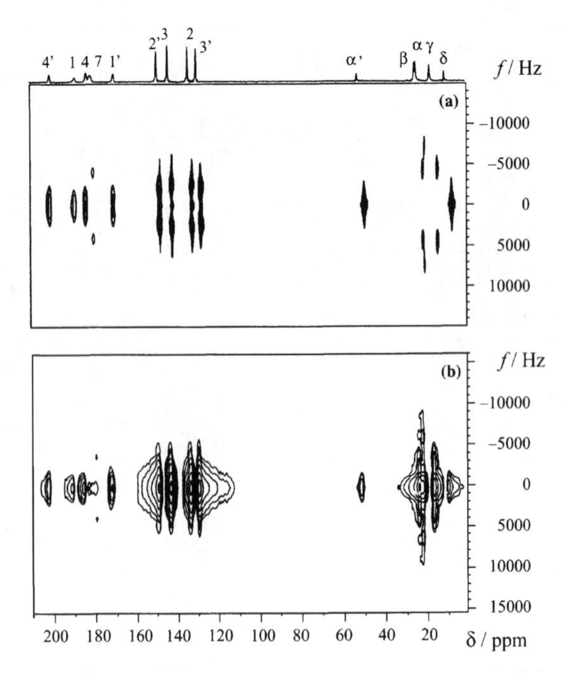

Fig. 8. ^{13}C SLF-2D NMR spectra of *MBBA* in its nematic phase obtained with (a) polarisation inversion and (b) standard cross polarisation pulse schemes on a Bruker DSX-300 NMR spectrometer at room temperature

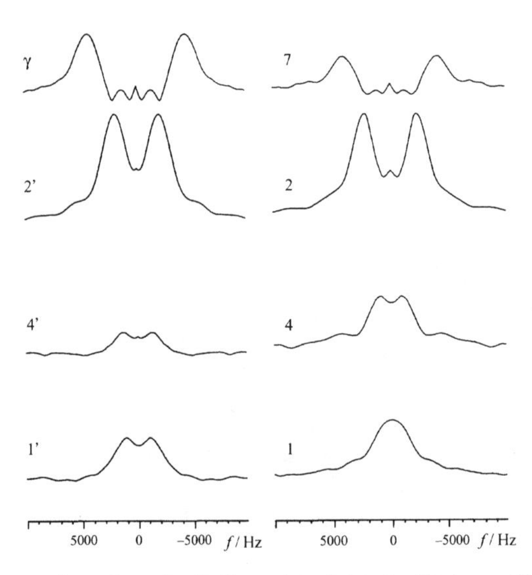

Fig. 9. Plots of cross-sections along the F_1 dimension giving the dipolar oscillation frequencies for γ, 7, 2′, 2, 4′, 4, 1′, and 1 carbons of *MBBA* obtained from the 2D plot shown in Fig. 8a

Cross-Polarisation from the Dipolar Reservoir under Mis-Matched *Hartmann–Hahn* Condition

In the studies reported so far, polarisation transfer has been considered between the *Zeeman* reservoirs of *I* and *S* spins under *Hartmann–Hahn* match. For a mismatch of the *Hartmann–Hahn* condition, an absorptive Lorentzian behavior for the *S* spin intensity has been predicted [19] with the width of the Lorentzian being related to the homonuclear dipolar coupling of the abundant spins. A similar study in which the initial magnetization is in the dipolar bath rather than in the *I* spin *Zeeman* bath has also been reported [20] and on the basis of the quasi-equilibrium theory a dispersive Lorentzian behaviour for the *S* spin intensity has been predicted. The above polarisation transfer process has been considered in detail by carrying out experimental measurements for a range of mismatch conditions by utilizing the pulse scheme shown in Fig. 1e [21]. Here, an ADRF pulse sequence on the *I* spins creates a dipolar order from the *I* spin *Zeeman* order during the period t_a. During t_{CP} a *Hartmann–Hahn* cross-polarisation pulse sequence is used, which results in the transfer of polarisation from the dipolar bath to the *S* spin *Zeeman* bath. The transfer process has the characteristics that *S* spin intensity is zero for $\omega_{1S}=\omega_{1I}$, positive for $\omega_{1S}>\omega_{1I}$ and negative for $\omega_{1S}<\omega_{1I}$ where ω_{1S} and ω_{1I} are the strengths of the spin-lock r.f. on the *S* and *I* spins respectively during the CP process. The *S* spin intensity plotted as a function of $\Delta\omega=\omega_{1S}-\omega_{1I}$ can be shown to be given by

$$M_S=\frac{\beta\lambda^2\Delta\omega}{\Delta\omega^2+\lambda^2}, \tag{7}$$

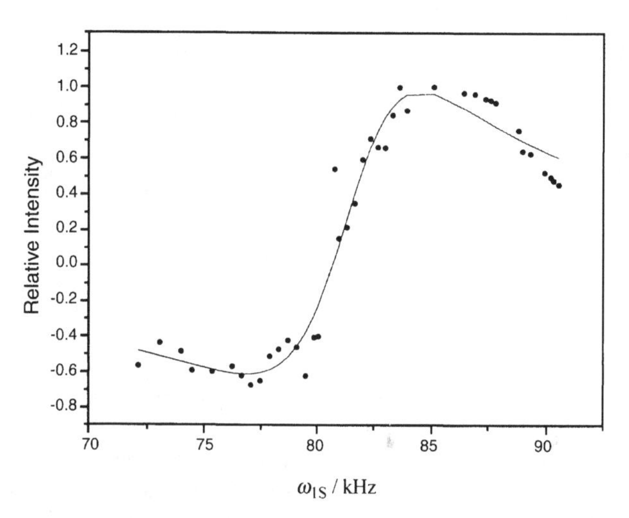

Fig. 10. Variation of cross-polarised signal intensity from the dipolar bath of protons to the α-carbons in the liquid crystal *EBBA* for different r.f. powers (ω_{1S}) on ^{13}C. The experimental points have been obtained by using the pulse sequence of Fig. 1e, with $t_a=t_b=1$ ms and $t_{CP}=200$ µsec. The spectra were obtained on a Bruker DSX-300 NMR spectrometer at a ^{13}C resonance frequency of 75.43 MHz. The proton r.f. power (ω_{1I}) was kept constant at 83 kHz. The continuous curve corresponds to a fit of the experimental data to a mixture of dispersive and absorptive Lorentzian functions and yields a value of $\lambda^2=14.2\,\text{kHz}^2$ for the α carbon

where β is the initial inverse spin temperature of the dipolar reservoir and λ^2 is related to the *I* spin second moment. The experimental results for the α carbon of *EBBA* in its oriented phase obtained by observing the ^{13}C resonance and by using the pulse sequence of Fig. 1e are shown in Fig. 10. The points correspond to the intensity of the carbon line as a function of ω_{1S} for a fixed value of ω_{1I}.

It is observed that the plot indeed shows an overall dispersive Lorentzian behaviour. However, the fit of the data to Eq. (7) showed some deviations. A better fit to the experimental data could be obtained by including an additional adsorptive Lorentzian of the same width which could arise due to transfer from remnant *Zeeman* magnetization of the *I* spins at the end of the ADRF period, cross polarising directly to the *S* spins [21]. This continuous curve shown in Fig. 10 corresponds to the fit obtained using the above procedure. A value 14.2 kHz^2 for λ^2 has been obtained for the α carbon of *EBBA*. Other methylene carbons along the chain show a variation of λ^2 that is expected on the basis of the variation of the local order parameters of the system. This method like the WISE technique [22], provides a means of indirectly monitoring proton dipolar couplings using the heteronuclear chemical shift dispersion.

Applications

The cross-polarisation techniques mentioned above have been utilized to obtain dipolar couplings and order parameters in several novel liquid crystalline systems such as:

(i) Aromatic systems containing four rings in the main core, a lateral hexyloxy chain, and a lateral aromatic branch with the aromatic ring itself being modified by different substitutents at *meta* or *para* position [23].
(ii) Systems containing four rings in the main core, one terminal, and two nearby lateral chains on each of the outer aromatic rings [24].
(iii) Molecules containing the 2-phenylindazole core – the first bonds of the two terminal chains on the either side are not along the same axis due to the presence of the five membered ring in the core [25].

As an illustration of the use of the C–H bond order parameter obtained from utilizing the cross polarisation technique, systems containing three aromatic rings with a lateral crown-ether fragment and with oxyethylene (*OE*) units replacing the terminal alkoxy chains is presented [26] here. The replacement of the alkoxy chains by chains containing the oxyethylene units decreases the melting and clearing temperature so as to obtain nematic compounds near room temperature. The symmetric mesogen containing one *OE* unit in the terminal chain, referred to as *CINPOE*1*Bu* (Fig. 11) has been studied. The ^{13}C chemical shifts in the isotropic phase and in the nematic phase at different temperatures have been monitored. The SLF 2D spectrum of the compound in its nematic phase at 349 K obtained using the pulse sequence of Fig. 1d is shown in Fig. 11. From the peaks in dipolar dimension, the magnitude of the dipolar interaction for several carbons have been obtained.

For example, carbon C_a and C_c have the same chemical shifts, but have different dipolar couplings, the larger one being attributed to C_a, the first carbon in the terminal chain. In the crown-ether segment, it is noticed that the dipolar couplings decreases from C_α to C_γ, but increases for C_δ. This indicates that the average angle

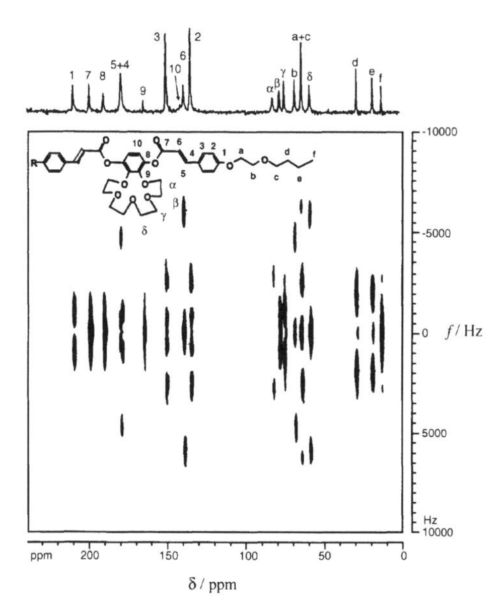

Fig. 11. [13]C 2D spectrum of *ClNPOE1Bu*, a liquid crystal containing a crown-ether and an oxyethylene unit in its nematic phase at 349 K, recorded on a Bruker DSX-300 NMR spectrometer at 75.47 MHz

between the C–H bond and the long molecular axis is less than the magic angle for carbons, C_α, nearer to the magic angle for C_β and C_γ and greater than magic angle for C_δ. This conclusion is supported by molecular modelling studies of the central part of a single molecule. For carbons in the polyoxyethylene chain, it is expected that the order parameter decreases monotonically with increasing distance from the core. This is in contrast to the odd–even effect usually observed in a terminal alkyl or alkoxy chain. Such a difference arises due to different probabilities for the conformers of the *POE* chain [27, 28]. In the system under consideration with one *OE* unit and one butyl fragment, the dipolar couplings decreases monotonically over the *OE* segment from carbon *a* to *c* while for the remainder of the carbons (*d* to *f*) the odd–even effect is present. The variation of the bond order parameter with temperature shows interesting correlation with the corresponding chemical shift of the carbon. A linear correlation between chemical shift and order parameter indicates that the conformation does not change significantly in the temperature range studied. Such a linear correlation has been observed for the carbons of the lateral crown-ether. On the other hand, the non-linear correlation observed for the carbons of the *OE* unit indicates that the conformational probabilities change significantly with temperature resulting in different averaged values of the dipolar couplings and chemical shift anisotropies.

Conclusions

The coherent effects of the cross-polarisation have been shown to provide detailed information on dipolar couplings in nematic liquid crystalline systems. The information thus obtained enables estimation of the order parameters of the liquid crystals. Several variants of the standard cross polarisation experiments were presented and their utility discussed. Typical applications of the method were presented. In addition to the above studies where heteronuclear dipolar couplings were used, another cross-polarisation method that can provide information about the proton–proton dipolar couplings was examined, thus providing another means of obtaining local order.

Acknowledgement

The authors would like to thank Prof. Anil Kumar for useful suggestions.

References

[1] Diehl P, Khetrapal CL (1969) NMR: Basic Principles and Progress **1**: 1
[2] Emsley JW (1985) Nucl Magn Reson Liq Cryst. Reidel, Dordrecht
[3] Dong RY (1994) Nucl Magn Reson Liq Cryst. Springer, New York
[4] Hohener A, Muller A, Ernst RR (1979) Mol Phys **38**: 909
[5] Magnuson ML, Tanner LF, Fung BM (1994) Liq Cryst **16**: 857
[6] Magnuson ML, Fung BM, Schadt M (1995) Liq Cryst **19**: 333
[7] Courtieu J, Bayle JP, Fung BM (1994) Prog Nucl Magn Reson Spectrosc **26**: 141
[8] Fung BM (1996) Encyclopedia of NMR **4**: 2744
[9] Caldarelli S, Hong M, Emsley L, Pines A (1996) J Phys Chem **100**: 18696
[10] Hong M, Pines A, Caldarelli S (1996) J Phys Chem **100**: 14815
[11] Caldarelli S, Lesage A, Emsley L (1996) J Am Chem Soc **118**: 12224
[12] Muller L, Anil Kumar, Baumann T, Ernst RR (1974) Phys Rev Lett **32**: 1402
[13] Hester RK, Ackerman JL, Cross VR, Waugh JS (1975) Phys Rev Lett **34**: 993
[14] Pratima R, Ramanathan KV (1996) J Magn Reson **A118**: 7
[15] Nagaraja CS, Ramanathan KV (1999) Liq Cryst **26**: 17
[16] Neeraj Sinha, Ramanathan KV (2000) Chem Phys Lett **332**: 125
[17] Wu C, Ramamoorthy A, Opella SJ (1994) J Magn Reson **A109**: 270
[18] Lee M, Goldburg WI (1965) Phys Rev **A140**: 1261
[19] Levitt MH, Suter D, Ernst RR (1986) J Chem Phys **84**: 4243
[20] Zhang S, Stejskal EO, Fornes RE, Wu X (1993) J Magn Reson **A104**: 177
[21] Venkatraman TN, Neeraj Sinha, Ramanathan KV (2002) J Magn Reson **157**: 137
[22] Schmidt-Rohr K, Clauss J, Spiess HW (1992) Macromolecules **25**: 3273
[23] Berdague P, Bayle JP, Fujimori H, Miyajima S (1998) New J Chem 1005
[24] Berdague P, Munier M, Judeinstein P, Bayle JP, Nagaraja CS, Ramanathan KV (1999) Liq Cryst **26**: 211
[25] Berdague P, Judeinstein P, Bayle JP, Nagaraja CS, Neeraj Sinha, Ramanathan KV (2001) Liq Cryst **28**: 197
[26] Neeraj Sinha, Ramanathan KV, Berdague P, Judeinstein P, Bayle JP (2002) Liq Cryst **29**: 449
[27] Rayssac V, Judeinstein P, Bayle JP, Kuwahara D, Ogata H, Miyajima S (1998) Liq Cryst **25**: 427
[28] Samulski ET, Dong RY (1982) J Chem Phys **77**: 5090

Improved Proton Decoupling in NMR Spectroscopy of Crystalline Solids Using the SPINAL-64 Sequence

Thomas Bräuniger*, **Philip Wormald**, and **Paul Hodgkinson**

Department of Chemistry, University of Durham, Durham DH1 3LE, United Kingdom

Received May 6, 2002; accepted May 22, 2002
Published online November 7, 2002

Summary. The performance of three different spin decoupling schemes, CW, TPPM, and SPINAL-64 is compared, by recording proton decoupled ^{13}C NMR spectra of a crystalline glycine sample, with 20% isotopic labelling. At a magnetic field of $B_0 = 14.1$ T, the two phase modulated pulse schemes, TPPM and SPINAL-64, are shown to give decisively better results than CW decoupling, and the SPINAL-64 sequence is found to perform better than TPPM. It is suggested that in NMR of crystalline solids, SPINAL-64 offers a viable and competitive alternative to the well established TPPM decoupling technique, especially at higher magnetic fields.

Keywords. Glycine; NMR spectroscopy; Proton decoupling; Solid state; SPINAL-64.

Introduction

Spin decoupling is one of the most important techniques in Nuclear Magnetic Resonance (NMR) spectroscopy, because it allows the acquisition of highly resolved and simplified spectra. In NMR, spectra of rare spins such as ^{13}C are usually observed with simultaneous decoupling of the abundant spins, which are most often protons, ^{1}H. This is done to remove spin–spin interactions (J couplings in liquid-state, heteronuclear dipolar interactions in solid-state), which otherwise might have deleterious effects on the resolution of the ^{13}C spectra.

The most straightforward approach to decoupling is continuous-wave (CW) irradiation at the resonance frequency of the target nuclei. In liquid-state NMR, CW decoupling with a sufficiently strong radio frequency (RF) field can give satisfactory results. However, much better decoupling performances are obtained by purpose-designed pulse sequences such as MLEV-4 [1], WALTZ-16 [2] or PAR-75 [3], which are less sensitive to offsets of the decoupler frequency. This robustness with respect to the frequency offset is one major criterion to judge the performance

---

* Corresponding author. E-mail: thomas.braeuniger@durham.ac.uk

of a decoupling technique, together with the desire to obtain good results for low RF power.

As opposed to liquid-state NMR, where decoupling aims to remove the relatively small scalar J couplings, the dominating effect in solid-state NMR is the dipolar interaction, which is generally comparable or even larger than the chemical shift range. For solids, much stronger RF fields are needed for CW decoupling, and the 'broad-band' pulse sequences devised for liquid-state NMR [1–3] are found mostly inefficient. In addition, the time dependence imposed on the proton resonance frequencies by the routinely used Magic Angle Spinning (MAS) technique [4] makes matters more complex. For rigid (crystalline) samples at modest MAS speeds and moderate magnetic fields, spin diffusion is still efficient enough for the proton lineshape to be largely unresolved. If the transmitter frequency is set to the middle of this broad ^{1}H line, CW usually gives better performance than other decoupling schemes. However, with the introduction of faster MAS speeds, and higher magnetic fields (and also for mobile samples), the proton lineshape starts to break up, rendering the spin diffusion process less efficient. This makes it essentially impossible to set the decoupler frequency precisely 'on resonance' [5], and decoupling performance tends to deteriorate.

The thus arising need for a decoupling technique more efficient than CW was addressed by the development of the Two Pulse Phase Modulated (TPPM) sequence [6]. This scheme relies on a windowless train of phase modulated pulses on the ^{1}H channel (see Fig. 1), and because of its significantly improved performance over CW decoupling, TPPM has found fairly widespread applications [7]. It has also been shown to be especially useful for triple-resonance experiments of the REDOR type [8], where the observed REDOR curves show an explicit dependence on the proton decoupler power [9], and converge to their 'true' shape only for high RF

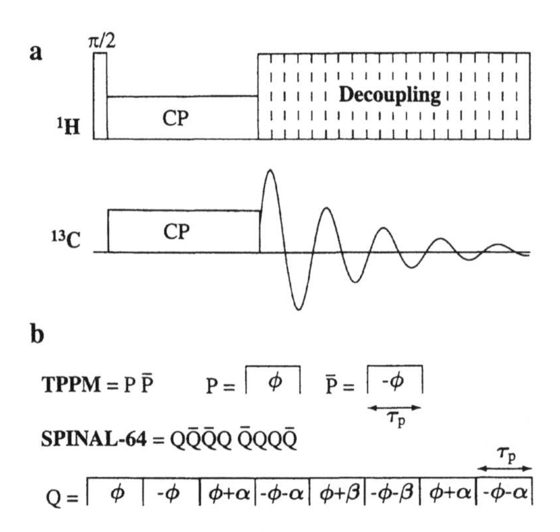

Fig. 1. Schematic representation of (a) the pulse sequence of the cross-polarization (CP) experiment, and (b) the TPPM and SPINAL-64 decoupling schemes, both using windowless pulse sequences with phase modulation on the ^{1}H channel. The phase angle ϕ, the pulse duration τ_p, and the phase increments α and β can be optimized for best performance

power. Since triple-resonance probes usually allow only for fairly limited RF field strengths, the implementation of decoupling schemes performing well at low RF power can improve REDOR results considerably [9, 10].

Following the success of TPPM, other decoupling schemes for solid-state NMR have been suggested [11–14]. The SPINAL-64 sequence [12] uses super-cycles of a basic phase modulated sequence (see Fig. 1). Although SPINAL-64 was shown to be a considerable improvement over TPPM in liquid crystalline systems, and to be somewhat better than TPPM and CW in at least one crystalline solid (*L*-tyrosine hydrochloride) [12], there is not much evidence of its use in the literature. Recently, however, the application of the SPINAL-64 sequence for proton decoupling of ^{13}C spectra of a fluorinated liquid crystal and a fluorinated alkane in an urea inclusion compounds has been reported [15]. In this study, SPINAL-64 gave better resolution than both CW and TPPM, the observed improvement was partly attributed to the high mobility of the decoupled species.

In this work, we demonstrate superior decoupling performance of the SPINAL-64 sequence at high magnetic field ($B_0 = 14.1$ T), for a crystalline sample of isotopically labelled glycine. In comparing the ^{13}C spectral linewidth obtained with CW, TPPM and SPINAL-64 decoupling, it is shown that SPINAL-64 produces the best line narrowing, and maintains good performance even at lower RF power levels. These results emphasize that the SPINAL-64 sequence can be successfully employed not only in systems with a relatively high mobility [12, 15], but also in rigid, crystalline samples.

Results and Discussion

The performance of the different decoupling schemes was tested on ^{13}C spectra of a glycine sample, which contained 20% of molecules isotopically labelled at both the α-carbon and the amino nitrogen, $^{15}NH_3^+{-}^{13}CH_2{-}COO^-$. The ^{13}C spectra of both labelled and unlabelled glycine are shown in Fig. 2a and 2b. For both samples, it can be seen that the α-carbon peak is conspicuously broader than the carbonyl resonance. For the unlabelled sample, a splitting of the α-carbon peak could be observed, which is most likely caused by the interaction of the ^{13}C spin with the neighbouring ^{14}N nucleus, which has spin $I = 1$ and a relatively large quadrupolar moment. This interaction cannot be the reason for the peak broadening seen in the 20% doubly labelled sample, because here practically all observed ^{13}C nuclei are bonded to ^{15}N nuclei with spin $I = 1/2$. For the labelled glycine, the carbonyl resonance broadens as well, although not to the extent that is seen for the α-carbon peak. One possible explanation of this effect is the presence of more than one crystalline modification in the sample. Three polymorphic forms are known to exist for crystalline glycine [16], and their slightly differing chemical shifts (amplified at the high magnetic field) could lead to inhomogeneous line broadening. However, for the purpose of this article, we will only focus on the effect of different decoupling methods on the ^{13}C spectrum of this sample, and not address the origin of the line broadening further.

The efficiency of the phase modulated decoupling sequences was optimized by iteratively testing arrays of the variable parameters, *i.e.* the phase angle ϕ, the pulse duration τ_p, and the additional phase increments α and β for SPINAL-64 (*cf.* Fig. 1). The original publications [6, 12] suggest $\tau_p \approx (11/12)\pi$, and the use of a small

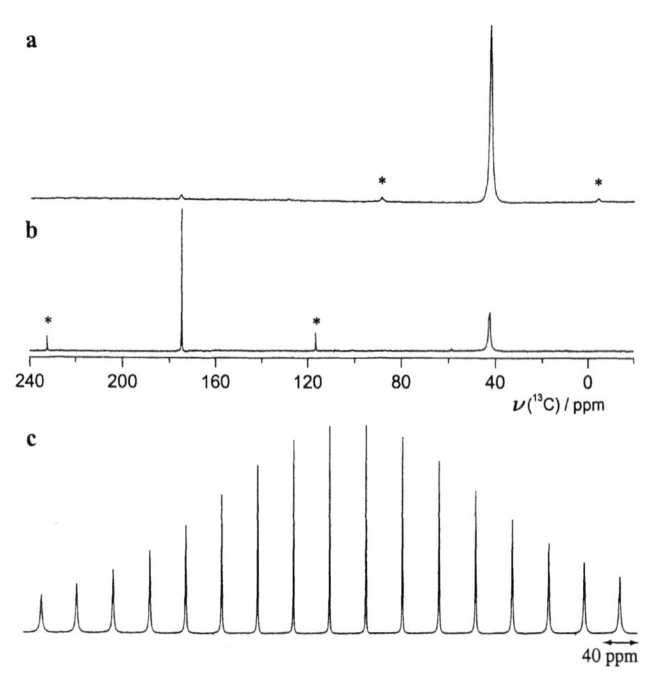

Fig. 2. (a) ^{13}C CP spectra with CW decoupling of (a) 20% doubly labelled glycine, at $\nu_0(^{13}C) = 150.9$ MHz, $\nu_r = 7.0$ kHz, and (b) unlabelled glycine, at $\nu_0(^{13}C) = 75.43$ MHz, $\nu_r = 4.4$ kHz. Asterisks indicate spinning side bands. (c) The dependence of the ^{13}C line shape (α-carbon peak at 42 ppm) in 20% labelled glycine on the decoupler offset (1 kHz stepwidth, 77 kHz RF power) for the SPINAL-64 sequence

angle for the phase modulation. For a decoupler strength of 77 kHz, we found the optimal values for τ_p to be $(11/12)\pi$ for SPINAL-64, and $(10/12)\pi$ for TPPM. Both sequences performed best with the originally reported phase values $\phi = 10°$ [6], and $\alpha = 5°$, $\beta = 10°$ [12].

The thus optimized sequences were first tested for their tolerance against the offset of the decoupler frequency. As an example, the line shapes obtained with SPINAL-64 are shown in Fig. 2c. The offset test was run for all three decoupling methods under consideration, and the full width at half maximum (FWHM) of the α-carbon peak determined as a measure of decoupling performance. The results are plotted in Fig. 3b. Clearly, both the phase modulated sequences outperform simple CW decoupling. It can also be seen that for frequency offsets up to ± 5 kHz, SPINAL-64 delivers the best line narrowing. Although TPPM produces less efficient decoupling in the ± 5 kHz offset range, its performance is slightly more robust than that of SPINAL-64 for very large frequency offsets.

In addition, we tested the tolerance of the different decoupling techniques with respect to the RF field strength applied to the ^{1}H channel. This was done deliberately without re-optimizing the pulse duration τ_p for the phase modulated sequences. Thus, with decreasing RF power, τ_p will increasingly deviate from the optimal value of $\approx (11/12)\pi$. As can be seen from Fig. 3a, SPINAL-64 shows

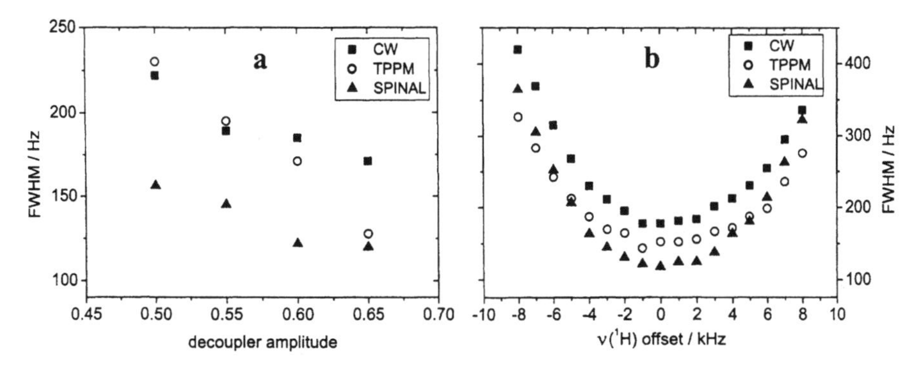

Fig. 3. Full Width at Half Maximum (FWHM) of the α-carbon peak of 20% doubly labelled glycine observed for CW, TPPM and SPINAL-64 decoupling. (a) Dependence on the decoupler amplitude, where 0.45 corresponds to an RF field strength of ≈ 54 kHz, and 0.65 to ≈ 77 kHz. (b) Dependence on the decoupler offset from the ideal value, $\nu(^1\text{H}) = 0$, acquired with a decoupler RF strength of ≈ 77 kHz

a remarkable tolerance to the decline in RF power, and is still performing well at low power levels. In contrast, TPPM displays a marked drop in decoupling efficiency, which can be restored again by using the τ_p values appropriate for the respective power settings.

Interestingly, we could not reproduce the same performance pattern when testing the decoupling sequences at the lower magnetic field ($B_0 = 7.05$ T). Here, all three decoupling methods gave similar results. Apparently, at this magnetic field and rotor spinning speed, proton spin diffusion is still efficient enough for simple CW decoupling to work well.

The reasons for the good performance of the SPINAL-64 sequence described here are not quite clear, as the physics of spin decoupling in solids is still only poorly understood. In general, decoupling of dipolar interactions has to cope with both dipolar and offset effects. SPINAL-64 was developed and optimized for liquid crystal systems, where molecular motion reduces the effective strength of the dipolar interactions. Since higher magnetic fields also tip the balance away from dipolar couplings towards chemical shift offsets, it is perhaps unsurprising that SPINAL-64 becomes more effective under these conditions.

In conclusion, both TPPM and SPINAL-64 offer decoupling performance superior to traditional CW irradiation, whenever experimental conditions (high B_0 and/or ν_r) lead to the breaking up of the broad ^{1}H lineshape and interfere with the proton spin diffusion process. The TPPM and SPINAL-64 sequences deliver quite robust decoupling when using the specified standard parameters, and require only minimal adjustments for optimal performance. In certain cases, SPINAL-64 may give the most efficient decoupling, as has been shown for highly mobile systems before [12, 15], and for relatively rigid crystalline samples in Ref. [12] and in this work. More experimental data and theoretical analysis on the application of this sequence are needed, but it seems that especially at high magnetic fields, SPINAL-64 can be considered a viable and competitive alternative to the well-established TPPM decoupling scheme.

Experimental

The doubly spin-labelled glycine, $^{15}NH_3^+$–$^{13}CH_2$–COO^-, was co-crystallized with the unlabelled compound to create a $\approx 20\%$ labelled sample. ^{13}C NMR spectra were acquired using cross-polarization (CP) with a contact time of 1 ms. All decoupler test spectra were run on a Chemagnetics Infinity spectrometer operating at $\nu_0(^{13}C) = 150.9$ MHz, and $\nu_0(^1H) = 600.11$ MHz, using a 4 mm Chemagnetics T3 probe, with a rotor spinning speed of $\nu_r = 7$ kHz. The spectrum of unlabelled glycine was acquired on a Varian INOVA spectrometer, operating at $\nu_0(^{13}C) = 75.43$ MHz, and $\nu_0(^1H) = 299.95$ MHz, using a Doty Scientific 7 mm probe, with a spinning speed of $\nu_r = 4.4$ kHz. The SPINAL-64 decoupling was conveniently programmed by a suitable modification of the Chemagnetics TPPM pulse program.

References

[1] Levitt MH, Freeman R (1981) J Magn Res **43**: 502
[2] Shaka AJ, Keeler J, Frenkiel T, Freeman R (1983) J Magn Res **52**: 335
[3] Fung BM (1984) J Magn Res **60**: 424
[4] Andrew ER (1981) Int Rev Phys Chem **1**: 195
[5] VanderHart DL, Campbell GC (1998) J Magn Res **134**: 88
[6] Bennett AE, Rienstra CM, Auger M, Lakshmi KV, Griffin RG (1995) J Chem Phys **103**: 6951
[7] See, for example, McGeorge G, Alderman DW, Grant DM (1998) J Magn Res **137**: 138
[8] Gullion T, Schaefer J (1989) J Magn Res **81**: 196
[9] Mitchell DJ, Evans JNS (1998) Chem Phys Lett **292**: 656
[10] Mehta AK, Hirsh DJ, Oyler N, Drobny GP, Schaefer J (2000) J Magn Res **145**: 156
[11] Gan ZH, Ernst RR (1997) Solid State NMR **8**: 153
[12] Fung BM, Khitrin AK, Ermolaev K (2000) J Magn Res **142**: 97
[13] Ernst M, Samoson A, Meier BH (2001) Chem Phys Lett **348**: 293
[14] Takegoshi K, Mizokami J, Terao T (2001) Chem Phys Lett **341**: 540
[15] Antonioli G, McMillan DE, Hodgkinson P (2001) Chem Phys Lett **344**: 68
[16] Perlovich GL, Hansen LK, Bauer-Brandl A (2001) J Therm Anal Calorim **66**: 699

The Flexibility of SIMPSON and SIMMOL for Numerical Simulations in Solid- and Liquid-State NMR Spectroscopy

Thomas Vosegaard*, **Anders Malmendal,** and **Niels C. Nielsen***

Interdisciplinary Nanoscience Center (iNANO) and Laboratory for Biomolecular NMR Spectroscopy, Department of Molecular and Structural Biology, University of Aarhus, DK-8000 Aarhus C, Denmark

Received June 27, 2002; accepted July 8, 2002
Published online November 7, 2002

Summary. Addressing the need for numerical simulations in the design and interpretation of advanced solid- and liquid-state NMR experiments, we present a number of novel features for numerical simulations based on the SIMPSON and SIMMOL open source software packages. Major attention is devoted to the flexibility of these Tcl-interfaced programs for numerical simulation of NMR experiments being complicated by demands for efficient powder averaging, large spin systems, and multiple-pulse rf irradiation. These features are exemplified by fast simulation of second-order quadrupolar powder patterns using crystallite interpolation, analysis of rotary resonance triple-quantum excitation for quadrupolar nuclei, iterative fitting of MQ-MAS spectra by combination of SIMPSON and MINUIT, simulation of multiple-dimensional PISEMA-type correlation experiments for macroscopically oriented membrane proteins, simulation of *Hartman-Hahn* polarization transfers in liquid-state NMR, and visualization of the spin evolution under complex composite broad-band excitation pulses.

Keywords. Solid-state NMR; Numerical simulations; Software; Membrane proteins; Inorganic materials.

Introduction

Over the past decades NMR spectroscopy has evolved tremendously by the development of powerful instrumentation and the design of thousands of advanced NMR experiments [1–3] offering almost complete control over the nuclear spin Hamiltonian to extract detailed information about the structure and dynamics of large molecules. This development has been facilitated by the use of theoretical tools such as average Hamiltonian theory [4–6] and the product operator formalism [7–9]. These tools have proved extremely useful for description of the spin

* Corresponding authors. E-mail: tv@chem.au.dk, ncn@imsb.au.dk

dynamics in relatively small spin systems subjected to well-defined rotations in pulsed experiments with a reasonable small number of distinct time events.

Parallel to the development of increasingly complex NMR experiments capable of handling more and more complex nuclear spin systems comes an increasing need for tools to accurately evaluate the performance of these experiments with the purpose of further development, artifact suppression, or interpretation of experimental data. This applies in particular for large spin systems, in presence of anisotropic nuclear spin interactions, and under conditions of imperfect rf irradiation. Solid-state NMR spectroscopy [10–14] very often faces all of these elements. Anisotropic nuclear spin interactions [11] introduce orientation dependent spin evolution which often is sufficiently strong that it not only affects free precession but also causes anisotropic evolution in periods with rf irradiation. In addition to this may come effects from ultra-fast sample spinning. Very often the experiments rely on delicate tailoring of the nuclear spin Hamiltonian by synchronous sample rotation and multiple-pulse rf irradiation to selectively re- and decouple specific interactions [15–27]. For example, methods of this kind have formed an indispensable basis for the current implementation of multiple-dimensional solid-state NMR for structural analysis of uniformly ^{13}C and ^{15}N labeled proteins immobilized by size, aggregation, or membrane association [28, 29].

To provide the spin engineers and application-oriented spectroscopists using NMR for structural analysis with tools for efficient numerical simulation of advanced solid- and liquid-state NMR experiments, we recently introduced the SIMPSON [30] and SIMMOL [31] programs as open source software. SIMPSON is a highly flexible program package for easy programming and simulation of essentially all types of solid-state NMR experiments. The amount and complexity of programming is very modest and similar to that required for implementation of pulse sequences on a commercial NMR spectrometers. As so this package may be considered a "computer spectrometer". To facilitate the setup of spin systems and in particular anisotropic interaction tensors for SIMPSON simulations, we recently introduced the SIMMOL program (and its predecessor PDB2SIMPSON [32]) to serve as a "sample-changer" for the "computer spectrometer". As an additional benefit, these programs enable straightforward visualization of relevant parts of the molecular structure with specification of relevant anisotropic NMR interactions.

Results and Discussion

While our previous accounts [30–32] have primarily described SIMPSON and SIMMOL individually as specific tools for solid-state NMR simulations, we will in this paper focus more on the flexibility of the two tools and their combination for numerical simulations in a broad spectrum of applications within solid- and liquid-state NMR spectroscopy. The key to this flexibility is the Tcl [33] scripting interfaces of SIMPSON and SIMMOL which offer unique possibilities for their control, mutual interaction, and even interaction with other programs through simple Tcl commands. These features will be demonstrated by (*i*) fast simulations of second-order quadrupolar powder patterns under magic-angle spinning (MAS) employing crystallite interpolation, (*ii*) fast simulation of triple-quantum (3Q) coherence excitation in quadrupolar nuclei using rotary resonance, (*iii*) fitting of experimental solid-state NMR spectra using SIMPSON in combination with MINUIT [34], (*iv*) SIMPSON and SIMMOL simulations of multi-dimensional PISEMA-type experiments for the study of macroscopically oriented membrane proteins, (*v*) optimization of *Hartman-Hahn* polarization transfer experiments in liquid-state NMR [3], and (*vi*) visualization of the spin evolution during complex hypersecant composite pulses. Overall these examples not only serve to demonstrate the flexibility of SIMPSON and SIMMOL but also represent important new extensions of these programs for advanced applications through development of routines for time-efficient calculations of complex state-of-the-art NMR experiments [35].

SIMPSON Flexibility: Fast Simulation of Second-Order Quadrupolar Powder Patterns by Crystallite Interpolation

Most of the nuclear spin interactions are orientation dependent. In solid-state NMR of poly-crystalline samples (and other samples with no orientational preference), the anisotropic interactions typically result in broad powder patterns because different orientations of the "crystallites" relative to the external magnetic field will be associated with different resonance frequencies [11, 14]. In such cases a simulated spectrum is achieved by integration over all possible crystallite orientations (i) each of which being characterized by three *Euler* angles ($\alpha_i, \beta_i, \gamma_i$) describing the orientation of the crystallite relative to the magnetic field. Since there are generally no analytical solutions to the integral, a standard procedure is to calculate the sum of the spectra for different crystallite orientations, weighted by the space angle spanned by the particular crystallite (w_i). Among the three *Euler* angles the γ-angle is special. For static powders the spectra are independent on γ, since the NMR interactions are invariant to rotation about the stationary field. For rotating solids, γ may be parametrized simultaneously with the sample rotation angle [36–40]. The α, β crystallites may be established randomly (Monte Carlo type), by rectangular or spherical grids, or more efficiently using tiling schemes with more uniform crystallite weight factors such as *Zaremba-Conroy-Wolfsberg* [41–43], SOPHE [44], *Alderman* [45], REPULSION [46], or Gaussian spherical quadrature [47] methods. While a relatively small set of well-defined *Euler* angles suffices for simulation of narrow powder patterns or spinning-sideband manifolds of such line shapes as being typical for MAS experiments of spin-$1/2$ nuclei, it is well-known that very broad powder patterns as often encountered for quadrupolar nuclei may require a much larger number of crystallite orientations in a carefully chosen tiling scheme.

Addressing exactly this problem, *Alderman et al.* [45] proposed an interpolation procedure that significantly reduces the number of crystallites needed to achieve a converged line shape. In this approach the resonance frequencies are first calculated for a relatively coarse set of α, β crystallites as represented by vertexes between the dark lines on the unit sphere in Fig. 1a. Next the resonance frequencies defining the vertexes of each triangle on the unit sphere (*e.g.*, the three crystallites labeled i, j, and k in Fig. 1a) are sorted as ν_{min}, ν_{mid}, and ν_{max}, representing the lowest, middle, and highest frequency, respectively. It is assumed that all crystallites within the space angle spanned by the triangle give resonance frequencies within the frequency range [ν_{min}, ν_{max}], and that the frequency spectrum

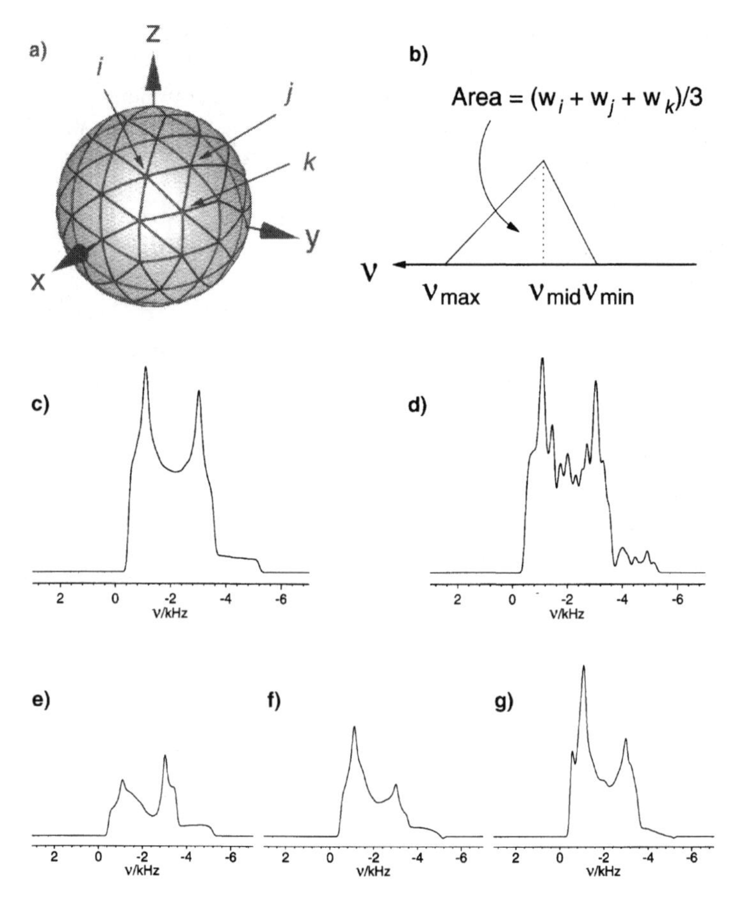

Fig. 1. Crystallite distribution of 100 REPULSION crystallites over the unit sphere. The crystallites labeled *i*, *j*, and *k* are located at the vertexes of a triangle. (b) Triangle corresponding to the spectrum from (α, β) points within the triangle defined by the crystallites *i*, *j*, and *k*. ν_{min}, ν_{mid}, and ν_{max} represent the lowest, middle, and highest resonance frequencies of the three vertexes. (c, d) Simulated 9.4-T ^{87}Rb powder spectra of $RbClO_4$ assuming ideal rf excitation, quadrupole coupling parameters of $C_Q = 3.3$ MHz and $\eta_Q = 0.21$ [50, 51], resulting from interpolation with 625 *Alderman* crystallites (c) and 678 REPULSION crystallites without interpolation (d). (e–g) Simulated 3Q-filtered MAS spectra with the same quadrupole coupling parameters and employing short-pulse (e), RIACT (f), and FASTER (g) 3Q excitation using the parameters in Table 1. All simulations (c–g) employ $\omega_r/2\pi = 30$ kHz and are apodized by a Gaussian line broadening of 150 Hz

for the triangle may be represented by a triangle ranging from ν_{min} to ν_{max} with maximum intensity at ν_{mid} as illustrated in Fig. 1b. The area of this triangle corresponds to the space angle of the crystallite triangle [45].

This interpolation method requires that the simulation can be performed directly in the frequency domain and that each crystallite gives a well-defined resonance frequency or set of such frequencies. This generally applies for static samples when the Hamiltonian is diagonal. For quadrupolar nuclei in MAS experiments, the second-order line shape is, to a good approximation, determined by a static component [48] depending only on the α and β angles (*vide infra*). Since γ and $\omega_r t$ both correspond to rotation around the rotor axis and the γ angle may distribute the intensity of the crystallite over the entire manifold of second-order quadrupolar spinning sidebands, this case is typically referred to as the infinite spinning frequency approximation. Although beyond the scope of this example, we note that interpolation-based powder averaging may also be constructed for more general cases with finite sample spinning [49] being particularly easy in the case of diagonal Hamiltonians [37]. In the typical case where the spectrum is excited using finite rf pulses, the orientation dependent intensity distribution may be included in the powder averaging by adjusting the weighting factor (w_i) for crystallite i to $w_i \rightarrow w_i \rho_i$, where ρ_i is the density operator element of interest for crystallite i at the beginning of the acquisition.

With the relevance of interpolation, in particular in presence of an orientation-dependent excitation profile, it is most relevant to address how this efficiently can be implemented into SIMPSON exploiting the flexibility of the Tcl interface. For the purpose of illustration, we consider the simulation of second-order quadrupolar powder patterns as achieved in the very popular multiple-quantum magic-angle spinning (MQ-MAS) experiments proposed by *Frydman* and co-workers [52, 53] for acquiring isotropic spectra of quadrupolar nuclei. This experiment relies on excitation of triple-quantum (3Q) coherences and mixing of these into detectable single-quantum coherences. Both of these processes are generally associated with rather low transfer amplitudes depending on the magnitude and orientation of the quadrupolar interaction tensor. For this reason the second-order quadrupolar powder pattern may be highly distorted by different excitation efficiencies for different crystallites. Such cases typically call for numerical simulations as demonstrated here by SIMPSON calculation of the 3Q excitation profile in MAS experiments. We note that the infinite spinning approximation may be justified in these examples since spinning sidebands are virtually non-existing in 3Q-filtered MAS experiments [54].

The resonance frequency for a single-quantum $(m, m-1)$ transition influenced by the second-order quadrupolar Hamiltonian ($\mathcal{H}_Q^{(2)}$) may be written

$$\begin{aligned}\omega_m(\alpha,\beta,\gamma;t) &= \langle m|\mathcal{H}_Q^{(2)}(\alpha,\beta,\gamma;t)|m\rangle - \langle m-1|\mathcal{H}_Q^{(2)}(\alpha,\beta,\gamma;t)|m-1\rangle \\ &= \sum_{l=-4}^{4} \omega_{m,m-1}^{l}(\alpha,\beta)\exp\{-il(\omega_R t+\gamma)\} \qquad (1)\end{aligned}$$

where the time-independent $l=0$ term governs the second-order line shape and $\omega_{m,m-1}^{0}$ is the resonance frequency for the α, β crystallite. The $l \neq 0$ terms are thereby responsible for the second-order quadrupolar spinning sidebands and are not relevant when operating in the infinite spinning rate approximation. We may isolate the $l=0$ term by averaging the resonance frequency over 2^n $(n \geq 3)$ γ crystallites which cancels all the oscillating terms.

In SIMPSON the Hamiltonian and density matrices are evaluated and stored in the memory during the calculation. These matrices may be accessed in the Tcl interface by the SIMPSON supplied commands `matrix get hamiltonian` and `matrix get density`. In a standard calculation of the solid-state NMR spectrum we will pick out the density operator element corresponding to I_- for the central $\left(\frac{1}{2},-\frac{1}{2}\right)$ transition (specified by the single-transition I_-^{2-3} operator for a spin-3/2 nucleus) and calculate the corresponding resonance frequency as described in Eq. (1). However, as we in this case merely want to investigate the impact of imperfect *excitation* (*i.e.*, assuming ideal 3Q → 1Q mixing) we will instead evaluate the density operator element corresponding to 3Q coherence (I_+^{1-4}) after the excitation. Once we know the density operator element and the resonance frequency for each set of crystallite *Euler* angles (α_i, β_i), the triangular interpolation may be performed as described above. It should be noted that this approach provides the exact spinning-frequency dependent excitation profile while the infinite spinning rate approximation is only used during the acquisition.

We have written a Tcl procedure to add the triangles to a spectrum (identified by the spectrum descriptor `$f`) as

```
faddtriangle $f $freq_i $freq_j $freq_k $w_re $w_im
```

which takes as input the resonance frequencies at the three vertexes (*i*, *j*, and *k*) in addition to the real and imaginary weighting factors for the triangle. The latter factors are calculated by Eqs. (2) and (3).

$$\mathtt{w\_re} = (w_i \mathrm{Re}(\rho_i) + w_j \mathrm{Re}(\rho_j) + w_k \mathrm{Re}(\rho_k))/3, \tag{2}$$

$$\mathtt{w\_im} = (w_i \mathrm{Im}(\rho_i) + w_j \mathrm{Im}(\rho_j) + w_k \mathrm{Im}(\rho_k))/3. \tag{3}$$

A first illustration of the advantage of using crystallite interpolation is given in Figs. 1c and 1d showing ideal (no effects from non-uniform excitation) second-order quadrupolar powder patterns for a ^{87}Rb nucleus with the quadrupole coupling parameters of $RbClO_4$ [50, 51] resulting from interpolation of 625 *Alderman* crystallites (Fig. 1c) and 678 REPULSION crystallites without interpolation (Fig. 1c). It is apparent from these spectra that far more crystallites are needed to achieve convergence of the line shape without interpolation, and that the result from the interpolation closely resembles an ideal second-order quadrupolar line shape. Different basis crystallite sets were used for the two simulations since we have observed that the uniformly distributed REPULSION crystallites typically provide better line shapes than *Alderman* crystallites when no interpolation is used, while the *Alderman* crystallites appear to be slightly better suited for interpolation.

Numerous techniques employing hard pulses [53], composite pulses [55], shaped pulses [56], rotation-induced adiabatic coherence transfer (RIACT) [57], fast amplitude-modulation [58–60], double frequency sweeps [61], and rotary-resonance (FASTER) [27] excitation have been presented in order to ameliorate the relatively low excitation and mixing efficiency and thereby sensitivity of the MQ-MAS experiment. We note that the sensitivity may additionally be improved by up to an order of magnitude in the detection part of the experiment by sampling through a train of QCPMG pulses [62, 63]. In the investigation of the powder patterns resulting from different excitation schemes, we will focus on three techniques, namely short-pulse, RIACT, and FASTER excitation. For each technique, we have performed a numerical search for the optimum parameters for the excitation pulses assuming ^{87}Rb quadrupole coupling parameters of $RbClO_4$ at 9.4 T and spinning at a frequency of 30 kHz. These results are listed in Table 1 and agree well

Table 1. Calculated optimum pulse lengths for the specified rf field strengths used for 3Q excitation with short-pulse, RIACT, and FASTER excitation schemes. The simulations employ $\omega_r/2\pi = 30\,\text{kHz}$ and the ^{87}Rb (spin $I = 3/2$) quadrupole coupling parameters for $RbClO_4$ ($C_Q = 3.3\,\text{MHz}$, $\eta_Q = 0.21$ [50, 51]) at 9.4 T

Excitation scheme	$\tau_p/\mu s$	$(\omega_{rf}/2\pi)/\text{kHz}$	Relative intensity
Short pulse	4.9	150	1
RIACT[a]	8.3	150	1.26
FASTER	99	37	1.78

[a] The RIACT pulse is preceded by a central-transition selective $\pi/2$ pulse employing an rf field strength of 40 kHz

with typical experimental findings. The simulated spectra obtained using interpolation of 625 *Alderman* crystallites are shown on the same intensity scale in Figs. 1e–1g for short-pulse (e), RIACT (f), and FASTER (g) excitation. Indeed the simulations reflect relative intensities in Table 1. As expected the simulated MQ-MAS line shapes are somewhat distorted compared to the ideal second-order quadrupolar line shape shown in Fig. 1c [63–66].

Overall, we have in this section demonstrated that time-efficient simulation of broad powder spectra using powder averaging in combination with crystallite interpolation may readily be implemented in SIMPSON. This approach has been demonstrated by simulation of second-order quadrupolar line shapes of 3Q filtered MQ-MAS experiments.

Fast Simulations with SIMPSON: Rotary-Resonance Triple-Quantum Excitation for Quadrupolar Nuclei

Analytical and numerical investigations of the effects of finite pulses on quadrupolar nuclei have been performed by numerous groups [27, 63, 67–70]. The prevailing approach has been to divide the calculation into small time increments over each of which the Hamiltonian is assumed to be constant, and evaluate the propagator for each of these fragments for a grid of (α, β, γ) crystallites [68]. As mentioned in the previous section, *Levitt* and *Edén* [36, 39], *Charpentier et al.* [38, 49], and *Hohwy et al.* [40] devised efficient methods for simulation of rotor-synchronous pulse sequences by performing the γ averaging simultaneously with the time incrementation. A comparison of these methods is beyond the scope of this paper, and we focus on the latter, so-called γ-COMPUTE algorithm [40], which is implemented in SIMPSON [30]. We note that γ-COMPUTE may be applied in simulations involving pulse sequences displaying appropriate periodicity with the rotor period [40].

The rotary-resonance [19, 71, 72] based FASTER [27] excitation scheme for MQ-MAS experiments discussed in the previous section reveals that for pulses longer than one rotor period, strong intensity of the 3Q coherence may be achieved when the rf field strength is in-between 3Q coherence zero-crossings at $\omega_{rf} = \omega_r \times n/2$ [27, 73, 74] with n being a positive integer. This effect becomes particularly pronounced at high spinning frequencies. The γ COMPUTE algorithm appears ideal

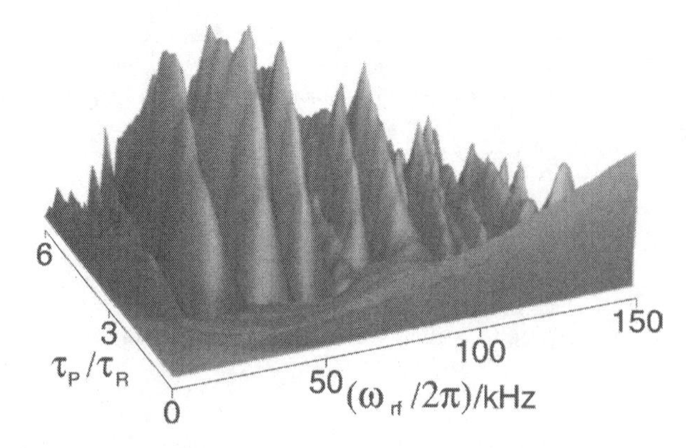

Fig. 2. SIMPSON simulated 3Q intensity as function of pulse length (τ_P) in units of rotor periods and the rf field strength under conditions of $\omega_r/2\pi = 30$ kHz spinning and quadrupole coupling parameters corresponding to ^{87}Rb in $RbClO_4$ [50, 51]

for simulation of this effect since the FASTER excitation sequence simply consists of an rf pulse with constant amplitude applied for several rotor periods.

For the purpose of illustration, we have performed a simulation of the 3Q coherence intensity as function of pulse length and rf field strength as shown in Fig. 2 employing a spinning frequency of 30 kHz and the ^{87}Rb quadrupole coupling parameters of $RbClO_4$ at 9.4 T with 50 time increments per rotor period for 6 rotor periods and rf field strengths from 0 to 150 kHz incremented in steps of 1 kHz. Employing 320 (α, β) REPULSION crystallites (and 50 γ angles) each rf trace was calculated within ~17 s using γ-COMPUTE, while the same result required ~450 s in a direct time-propagation simulation on a 1.9 GHz Pentium IV/Linux workstation. We note that the latter calculation may be speeded up significantly by calculating the propagators only for the first rotor period and subsequently reuse these propagators for the remaining five rotor periods employing the propagator storing facilities in SIMPSON. Using this approach, the calculation of each rf trace lasts ~177 s. In all simulations the time over which the Hamiltonian is assumed to be constant was 1/100 of a rotor period.

Overall, we demonstrated that time savings in the order of a factor 25 may be achieved using the γ-COMPUTE algorithm instead of direct propagator for simulation of 3Q coherence excitation profiles for FASTER experiments, and a ten-fold time saving as compared to direct propagation with reuse of the propagators. This example shows that although the advantage of γ-COMPUTE is most pronounced for large numbers of points extending over many rotor periods [40], it may still be beneficial to employ this procedure for pulses significantly shorter than one rotor period. In the present example the γ-COMPUTE algorithm remains faster than the direct propagation method for pulse lengths exceeding ~$0.2\tau_r$.

SIMPSON Optimization: Combination with MINUIT

To enhance the flexibility of SIMPSON and the capabilities for numerical optimization, we have extended SIMPSON with the minimization tools of the MINUIT minimization software package from CERN [34]. MINUIT represents a very powerful minimization tool providing options for Monte Carlo optimization, gradient and non-linear minimization, parameter scans, confidence interval calculation, *etc.*, while it in full analogy to SIMPSON and SIMMOL is open-source software. In a typical MINUIT minimization, the user has full flexibility to specify the parameters to vary during the minimization, the function to minimize, and the minimization procedure.

SIMPSON may be combined with MINUIT via a Tcl interface alternatingly accessing the two independently compiled programs [75] which demonstrates the great flexibility offered by scripting control. More elegantly, however, we here combine the two programs on the compiled level such that the MINUIT functions may be accessed directly from the SIMPSON Tcl interface. In this setup, the parameters to be optimized may be specified by the command

```
mnpar <name> <value> <error> <min> <max>
```

where the internal name of the parameter, its initial value, the estimated error, and optional minimum and maximum limits are specified. The value of this MINUIT parameter is stored in a Tcl variable `$mn(<name>)` and is thereby accessible from the Tcl interface. The function to minimize is a Tcl procedure `minuit` which returns the value subject to minimization. A typical scenario for determining the isotropic chemical shift and the quadrupolar coupling parameters (C_Q and η_Q) of an experimental spectrum (previously loaded into the descriptor `$g`) could be

```
proc minuit {} {
  global mn g # Make parameters globally available

  set f [fsimpson [list \
    [list shift_1_iso           $mn(iso)] \
    [list quadrupole_1_aniso $mn(cq)]  \
    [list quadrupole_1_eta    $mn(eta)] \
  ]]
  set rms [frms $f $g - re]
  funload $f
  return $rms
}
```

using various standard SIMPSON commands [30]. The minimization procedure is subsequently specified and may involve parameter scans, minimization, fixing and releasing parameters etc. All of these functions are available as SIMPSON Tcl functions, *e.g.*, `mnscan`, `mnminimize`, `mnfix`, *etc.*, and a typical minimization procedure for the above example could be

```
# Set up parameters
mnpar iso     52     5
mnpar cq       6.8   10
mnpar eta      0.08   0.1   01
```

```
# Minimization procedure
mnscan     iso
mnfix      eta
mnminimize
mnrelease eta
mnminimize
```

which will first perform a linear scan of the isotropic shift, minimize while fixing the quadrupole coupling asymmetry parameter, release it, and do the final minimization of all parameters.

To demonstrate this kind of minimization, we have investigated a previously published [76] ^{27}Al MQ-MAS spectrum of $9Al_2O_3 \cdot 2B_2O_3$ shown in Fig. 3a and recorded at 7.1 T with a spinning frequency of 15 kHz. The trace showing the quadrupolar line shape for the most intense of the Al_V sites is displayed in Fig. 3b. Apart from random noise this trace reveals a quite distorted second-order quadrupolar line shape due to imperfect 3Q excitation and mixing. We will use the interpolation scheme described in the previous section to simulate this spectrum including effects from finite pulses with the experimental values for the pulse lengths ($\tau_{exc} = 3.0\,\mu s$, $\tau_{mix} = 1.4\,\mu s$) and rf field strength ($\omega_{rf}/2\pi = 90$ kHz) [76]. Along with the previously reported quadrupole coupling parameters ($C_Q =$ 6.8 MHz, $\eta_Q = 0.08$, and $\delta_{iso} = 52$ ppm [76]), this should form an excellent starting point for a SIMPSON-MINUIT optimization of the quadrupole coupling parameters from an optimization procedure similar to the one sketched above. The parameters resulting from this optimization are $C_Q = 7.0 \pm 0.2$ MHz, $\eta_Q = 0.11 \pm 0.08$, and $\delta_{iso} = 56 \pm 2$ ppm [35] in good agreement with previously published parameters [76, 77]. The corresponding simulation is shown in Fig. 3c and does indeed reproduce most of the features of the experimental spectrum. For comparison, the

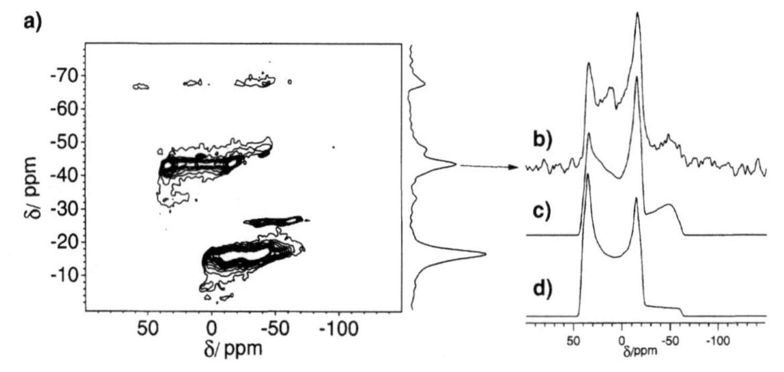

Fig. 3. (a) ^{27}Al MQ-MAS spectrum of $9Al_2O_3 \cdot 2B_2O_3$ recorded at 7.1 T ($\omega_0/2\pi = 78.2$ MHz) using $\omega_r/2\pi = 15$ kHz [76]. The spectrum is obtained using short-pulse excitation and mixing with an rf field strength of 90 kHz and pulse lengths of $\tau_{exc} = 3.0\,\mu s$ and $\tau_{mix} = 1.4\,\mu s$. (b) Trace showing the anisotropic line shape for the most intense Al_V site. (c) Simulation including finite-pulse effects and corresponding to the optimum quadrupole coupling parameters ($C_Q = 7.0$ MHz, $\eta_Q = 0.11$, $\delta_{iso} = 56$ ppm). (d) Same simulation but assuming ideal excitation and mixing

simulation resulting from ideal excitation but with identical quadrupole coupling parameters is shown in Fig. 3d. In contrast to the finite-pulse simulation the latter ideal simulation is incapable of modeling the intensity of the experimental spectrum.

To summarize, we have demonstrated that the Tcl controlled SIMPSON and SIMMOL programs can readily be combined with other programs or open source software packages. The specific example addresses the combination of SIMPSON with MINUIT which greatly enhances the capabilities for numerical minimizations for extraction of accurate parameters from experimental NMR spectra or optimization of experimental procedures.

SIMMOL Simulations: Multi-Dimensional PISEMA-Type Experiments for Biological Solid-State NMR

One of the most successful protocols for solid-state NMR studies of membrane proteins, involves macroscopic orientation of the membranes (*e.g.*, phospholipid bilayers) in which the proteins are embedded on glass plates with the membrane normal parallel to the external magnetic field [28]. Since the nuclear spin interactions of the peptide backbone atoms largely possess the same magnitude and orientation relative to the peptide plane, independently on the secondary structure and residue type [31, 78–80], the observed resonance frequencies to a good approximation depend only on the overall orientation of the involved peptide plane relative to the magnetic field direction and thereby to the lipid bilayers. Consequently, multi-dimensional NMR experiments of oriented proteins will display unique resonance patterns depending on the peptide secondary structure and its conformation in the membrane. The use of such topology-aided structure elucidation has recently been demonstrated for uniformly ^{15}N-labelled α-helical peptides [80, 81] which lead to characteristic circular resonance patterns – the so-called polarization index slant angle (PISA) wheels – in 2D ^{1}H–^{15}N dipolar coupling/^{15}N shift (PISEMA) correlation experiments [82]. Obviously, the PISA wheels are not unique for the interactions involved in 2D PISEMA experiments as recently illustrated for 2D ^{1}H/^{15}N chemical-shift correlation experiments [83] and a variety of different 2D double- and triple-resonance experiments [84].

A key feature of SIMMOL is its ability to associate different atoms in protein structures (*e.g.*, structures downloaded from the protein database [85] or synthetic structures created using SIMMOL or programs such as WHAT IF [86]) with typical parameters for the magnitude and orientation of anisotropic nuclear spin interaction tensors. This makes SIMMOL an extremely powerful tool for simulation of biological solid-state NMR experiments, such as those described above, on real structures. The SIMMOL output setting up the correct magnitude and *Euler* angles for all specified interactions may be directed to the spin-system part of the SIMPSON input file and subsequently used in a SIMPSON simulation taking into account all ideal/non-ideal features of the NMR experiment as demonstrated elsewhere [31, 32, 84]. However, for fast simulation of PISA wheels we may also choose to disregard the effects from finite pulses and calculate the multiple-dimensional resonance frequencies for the different residues directly in SIMMOL [31]. To do this we use the Tcl output from the SIMMOL commands assigning tensorial interactions to specific atoms, e.g., `mshift <buffer1>` which specifies chemical shift tensors for the atoms previously assigned to buffer 1 or `mdipole <buffer1> <buffer2> <max> <min>` which identifies all dipole-dipole couplings between atoms in buffer 1 and 2 in the range from `max` to `min`, specified as frequencies or distances. These commands return the atom numbers (from the PDB file) of the involved atoms, the magnitude for the interaction (*e.g.*, δ_{iso}, δ_{aniso}, and η), and the *Euler* angles (Ω_{PL}) describing the orientation of the tensor relative to the laboratory frame.

With the magnetic field oriented along z in the laboratory frame (L), the resonance frequency for an anisotropic interaction may be calculated by transferring the description of the tensor from its principal axis system (P) to L by the transformation

$$A_L = R(\Omega_{PL})A_P R(\Omega_{PL})^T \tag{4}$$

with A_P and $R(\{\Omega\}_{PL})$ denoting the diagonal principal axis system tensor and the Cartesian rotation matrix. The resonance frequency corresponding to any first-order Hamiltonian [4] is given by the zz element of the A_L tensor. We have implemented this coordinate transformation in a Tcl procedure `zzlab <a_p> <angles>`, which uses the SIMMOL Tcl commands `mgeteulermatrix` and `mmath` [31] to manipulate the matrices.

For calculation of PISA wheels, we may use a SIMMOL-synthesized peptide with torsion angles corresponding to ideal α-helix ($\phi = -65°$, $\psi = -40°$), β-strand ($\phi = -135°$, $\psi = 140°$), or any other values of interest and rotate this to form a given tilt angle relative to the magnetic field z axis. Equipped with these structures, SIMMOL allows easy generation of PISA wheels by calculating the resonance frequencies as described above for small increments of the rotational pitch of the peptide. Figure 4 contains four examples of resonance wheel patterns obtained for ideal α-helices tilted differently with respect to the magnetic field and corresponding to a variety of 2D and 3D ^{1}H–^{15}N dipolar coupling, ^{15}N shift, and ^{13}C′ shift correlation experiments. The particular experiments in Fig. 4 are chosen since they have recently been highlighted as particularly relevant for large proteins since they have very attractive properties with respect to resolution power and assignment feasibility [84].

With these simulations we have demonstrated the versatility of SIMMOL for assigning orientation-dependent nuclear spin interactions to polypeptide structures and, with proper Tcl procedures, directly simulate multi-dimensional NMR experiments for uniaxially oriented polypeptides.

SIMPSON and SIMMOL: Optimization of Homonuclear Hartmann-Hahn Polarization Transfer in Liquid-State NMR

The need for reliable and flexible spectral simulations is by no means restricted to solid-state NMR. Although many of the liquid-state applications are performed with standard parameters or are optimized automatically these days, a robust simulation and optimization tool comes in handy when taking even a small step away form the well-paved main road.

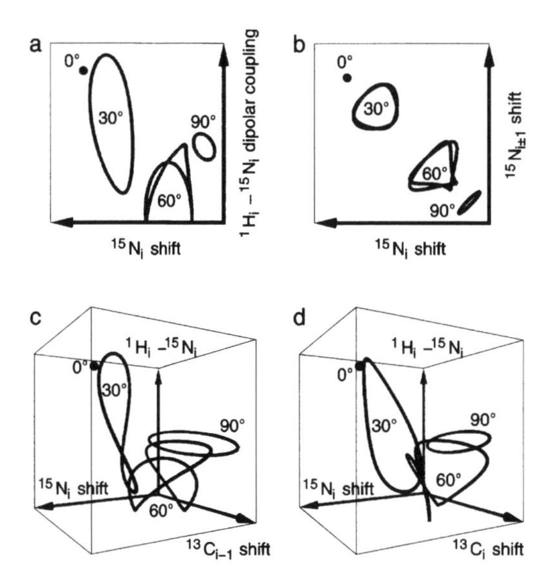

Fig. 4. SIMMOL simulated 2D (a, b) and 3D (c, d) PISA wheels for an ideal α-helix ($\phi = -65°$, $\psi = -40°$) structure with tilt angles of 0°, 30°, 60°, and 90° relative to the magnetic field direction. The simulations represent (a) ^1H_i–$^{15}N_i$ dipolar coupling/$^{15}N_i$ shift (PISEMA), (b) $^{15}N_i/^{15}N_{i\,\pm\,1}$ shift, (c) ^1H_i–$^{15}N_i$ dipolar coupling/$^{15}N_i$ shift/$^{13}C'_{i-1}$ shift, and ^1H_i–$^{15}N_i$ dipolar coupling/$^{15}N_i$ shift/$^{13}C'_i$ shift correlation experiments. The displayed frequency ranges (with the arrows pointing from lower to higher frequency) are 0–10 kHz for the ^1H–^{15}N dipolar coupling dimension, 50–250 ppm for ^{15}N shift dimensions, and 75–250 ppm for the $^{13}C'$ shift dimensions

An example of parameters that may deserve some attention is that associated with *Hartmann-Hahn* polarization transfer, either it is used for non-selective or selective, homo- or hetero-nuclear transfer. Specifically, we will here look at the possibilities for homonuclear transfer of polarization from the C^α to the outermost aliphatic carbon (C^β or C^γ) in Asx or Glx residues (asparagine and aspartic acid, or glutamine and glutamic acid) at 18.8 T with maximal transfer amplitudes. For the purpose of illustration, we examine the transfer efficiency of DIPSI-2 [87] as function of the rf field strength ($\omega_{\text{rf}}/2\pi$), the mixing time (τ_{mix}), and the carrier frequency ($\omega_{\text{offset}}/2\pi$).

The low dimensionality of the problem makes it amenable to manual optimization using grid plots as illustrated in Fig. 5. Alternatively, it is straightforward to perform a real optimization using SIMPSON and MINUIT as described in the previous section. Two-dimensional contour plots were calculated for the amplitudes of Asx $C^\alpha \rightarrow C^\beta$ and Glx $C^\alpha \rightarrow C^\gamma$ transfer as a function of τ_{mix} and $\omega_{\text{offset}}/2\pi$ at a constant rf field strength of 10 kHz (Fig. 5a–b). Figure 5a shows that the Asx $C^\alpha \rightarrow C^\beta$ transfer agrees with the expected transfer for a two spin system, *i.e.*, optimal transfer at $\tau_{\text{mix}} = 1/(2J_{\text{CC}})$, $3/(2J_{\text{CC}})$, *etc.* when the carrier is placed in the middle of the spectrum. Although the transfer path is twice as long for Glx, the optimal τ_{mix} values are not much higher ($\sim 1.5/(2J_{\text{CC}})$, $\sim 3.4/(2J_{\text{CC}})$) implying that it should be possible to optimize the Asx and Glx transfers simultaneously. It is

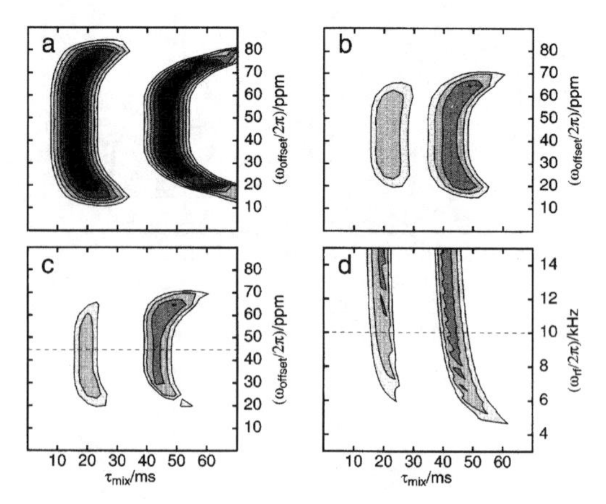

Fig. 5. SIMPSON simulated contour plots of (a) Asx $C^{\alpha} \rightarrow C^{\beta}$ polarization transfer, (b) Glx $C^{\alpha} \rightarrow C^{\gamma}$ transfer, and (c) min{Asx $C^{\alpha} \rightarrow C^{\beta}$, Glx $C^{\alpha} \rightarrow C^{\gamma}$} as a function of τ_{mix} and $\omega_{offset}/2\pi$ at $\omega_{rf}/2\pi = 10$ kHz, and (d) min{Asx $C^{\alpha} \rightarrow C^{\beta}$, Glx $C^{\alpha} \rightarrow C^{\gamma}$} as a function of τ_{mix} and $\omega_{rf}/2\pi$ at $\omega_{offset}/2\pi = 45$ ppm from TMS as indicated by a dashed line in (c). Likewise the rf field strength of 10 kHz employed in (a–c) is shown by a dashed line in (d). The contour levels mark 50% to 90% transfer efficiency in 10% steps, filled with increasingly dark gray. The shift parameters used in these simulations are for Asx: $\delta(C^{\alpha}) = 55$ ppm, $\delta(C^{\beta}) = 40$ ppm, $J_{CC} = 35$ Hz; and for Glx: $\delta(C^{\alpha}) = 55$ ppm, $\delta(C^{\beta}) = 30$ ppm, $\delta(C^{\gamma}) = 35$ ppm. $J_{CC} = 35$ Hz in all cases [88, 89]

interesting to note, that the Glx $C^{\alpha} \rightarrow C^{\gamma}$ transfer tends to increase when the carrier is placed in the region above $\delta(C^{\alpha})$.

To search for optimal sensitivity, contour plots where at each point the smallest of the two transfer amplitudes (min{Asx $C^{\alpha} \rightarrow C^{\beta}$, Glx $C^{\alpha} \rightarrow C^{\gamma}$}) was plotted as a function of τ_{mix} and $\omega_{offset}/2\pi$ at constant $\omega_{rf}/2\pi$ (Fig. 5c), and as a function of τ_{mix} and $\omega_{rf}/2\pi$ at constant $\omega_{offset}/2\pi$ (Fig. 5d) were calculated. These plots show that a transfer of at least 60% for both residue types can be obtained at $\tau_{mix} = 24$ ms ($\sim 1.7/2J_{CC}$) and $\omega_{rf}/2\pi = 7.3$ kHz and 70% at $\tau_{mix} = 21$ ms ($\sim 1.5/2J_{CC}$) and $\omega_{rf}/2\pi = 11$ kHz. If τ_{mix} is increased to 48 ms, a further gain in sensitivity can be achieved, or $\omega_{rf}/2\pi$ can be reduced. However, these gains will compete with losses due to relaxation effects. These were left out for simplicity and will indeed be the limiting factor for larger molecules.

When the backbone carbonyl (C′) and sidechain carbonyls and carboxylates (C^{γ} and C^{δ} for Asx and Glx, respectively) are included in the simulations (not shown) we observe that the maximum efficiency for the desired transfer to the outermost *aliphatic* carbon decreases slightly when the either the rf field strength or the carrier offset are increased. However, the contour plots are virtually identical to those presented in Fig. 5.

SIMPSON and SIMMOL: Visualization of the Spin Evolution During Pulse Sequences

With todays large magnetic field strengths, the bandwidth of normal square pulses of reasonable power may in many cases be too small to cover the whole spectrum of interest. For example, a ^{13}C spectrum spans about 40 kHz at 18.8 T. Thus, it is of great interest to construct pulses that behave uniformly over this frequency range. *Pines* and co-workers [90, 91] demonstrated that pulses with hyperbolic secant (HS) phase shifts had nice broadband inversion properties, an observation that has later been pursued by several other groups [92–94]. Recently, *Shaka* and co-workers [95] described the use of composite HS pulses for broadband excitation. In particular they demonstrated how three consecutive HS pulses, $HS90^\circ_x(\tau) - HS180^\circ_y(\tau) - (\tau/2+\varepsilon) - HS180^\circ_y(\tau/2)$, may compensate the unfortunate phase and rf field inhomogeneity behavior of a single 90° HS pulse.

While it is generally accepted that the performance of complicated pulses may exceed that of the standard rectangular pulses, it is much more difficult to visualize and evaluate these pulses and their impact on the spin evolution. In this section we will demonstrate that SIMPSON and SIMMOL through their flexibility allow such analysis to be performed without too much effort. For this purpose, we analyze the HS pulses of *Shaka* and co-workers [95] using the following functionality for the phase and amplitude of the pulses

$$\phi(t) = \phi(0) + \phi_0 \ln\left\{\cosh\left(\frac{10.6t}{\tau}\right)\right\} \tag{5}$$

$$\omega_{rf}(t) = \omega_{rf}^{max}/\cosh\left(\frac{10.6t}{\tau}\right) \tag{6}$$

with $\phi_0 = \pi f_{max}\tau/10.6$. τ and f_{max} represent the duration and the band width of the pulse. The three pulse shapes relevant for the composite HS pulse is visualized in Fig. 6a.

These rf schemes may straightforwardly be implemented into SIMPSON at the Tcl scripting level (*i.e.*, in the standard input file) by defining a new procedure `hspulse <tau> <wrfmax> <phase>` with a similar structure as the SIMPSON built-in procedure `pulse <tau> <wrf> <phase>`, *i.e.*,

```
proc hspulse {tau wrfmax phase} {
   set fmax 38e3    # Band width
   set dt   2.0     # Time increment
   set steps [expr int(round($tau/$dt/2))]
   set phi0 [expr 180.0*$fmax*$tau/10.6e6]

   for {set i -$steps} {$i < $steps} {incr i} {
     set x [expr cosh(10.6*$dt*$i/$tau)]
     set phi [expr $phase + $phi0*log($x)]
     set rf [expr $wrfmax/$x]
     pulse $dt $rf $phi
   }
}
```

which allows the three HS pulse sequence ($\tau = 4$ ms, $\varepsilon = 1.8\,\mu$s, and $\omega_{rf}^{max}/2\pi$ values of 1.88, 5.64, and 8.41 kHz) to be implemented as

```
hspulse 4000 1880 0
hspulse 4000 5460 90
delay   2001.8
hspulse 2000 8410 90
```

in a SIMPSON input file.

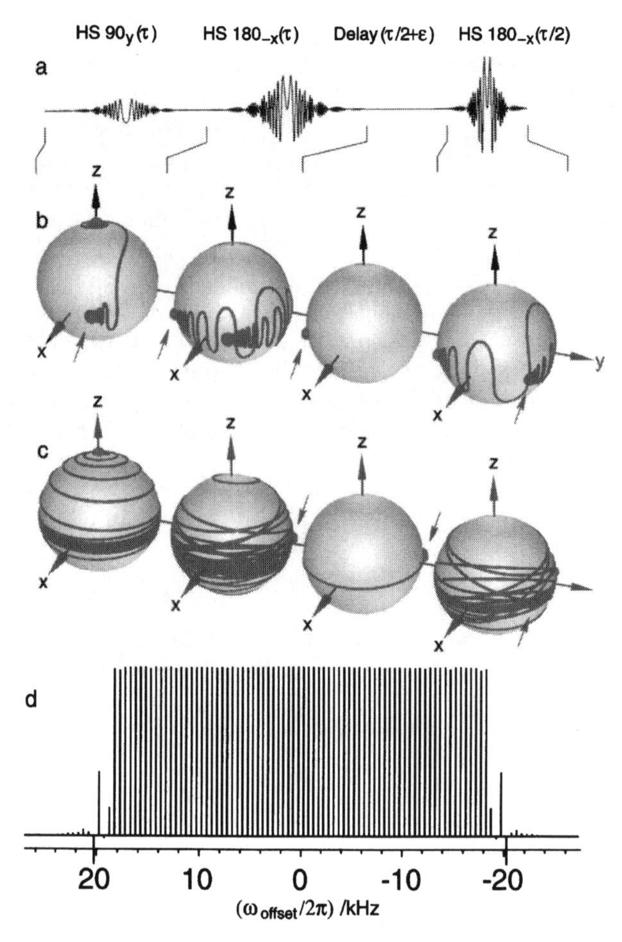

Fig. 6. Pulse-shape (a) employing a train of three hyperbolic pulses employed for broadband excitation [95] (see text). The SIMPSON simulated magnetization trajectories are shown in (b, c) for on-resonance pulses (b) or pulses with an offset of 15 kHz (c). Starting and ending points for the pulse sequence elements are highlighted by small spheres, and in addition the latter points are marked by small arrows. The band width of the pulse is set to $f_{max} = 38$ kHz. (d) Excitation profile as a function of $\omega_{offset}/2\pi$ for the pulse sequence in (a)

As a first illustration of the broadbandedness of the composite HS pulse Fig. 6d shows the offset dependence of the pulse as depicted by acquiring the excitation profile ($I_z \rightarrow I_-$) with different transmitter carrier offsets. This simulation employs the above parameters and corresponds to the sum of two experiments with all phases changed in order to achieve a symmetric excitation profile. Indeed, a band width approximately corresponding to f_{max} is achieved with a very uniform intensity. We note that the simulation in Fig. 6d compares favorably with the experimental result shown in Fig. 7 of Ref. [95].

To gain further insight in the magnetization trajectory during the composite pulse, we have performed a simulation where each of the spin operators I_x, I_y, and

I_z is acquired during the pulse. In SIMPSON this may be done by adding the commands

```
matrix set detect operator I1x
acq
matrix set detect operator I1y
acq
matrix set detect operator I1z
acq
```

in the loop of the `hspulse` procedure. Equipped with these three components, the magnetization trajectory may readily be visualized using any 3D visualization software. To illustrate the versatility of our tools, we have used the "non-molecular" visualization functions of SIMMOL (cylinders, arrows, and spheres) to create the images of the magnetization trajectories for on-resonance (Fig. 6b) and 15 kHz-off-resonance (Fig. 6c) pulses. In these figures the large spheres represent the unit sphere on which the magnetization vector travels when neglecting relaxation. The starting- and ending-points of each pulse sequence element are represented by small balls, and in addition the ending point is highlighted by a small arrow. The dark lines on the unit spheres show the magnetization excursions during the pulse sequence elements. We note that both the on-resonance and 15 kHz-off-resonance simulations indeed perform 90° rotations with a *xy* phase independent on the transmitter carrier offset in agreement with the reported properties of these composite pulses [95].

Conclusion

In conclusion, we have through a broad range of different NMR applications demonstrated that the flexibility of SIMPSON and SIMMOL – as provided through the Tcl user interfaces – allows for simulation of essentially all kinds of NMR experiments (disregarding effects from relaxation). This applies for direct simulation of NMR spectra as well as visualization and analysis of the inner working of complex pulse sequences. Besides demonstrating the versatility of SIMPSON and SIMMOL, the various examples have also been selected to add new and powerful features to these programs for the benefit of the users [35]. Beyond any discussion there is a tremendous, and steadily increasing, need for numerical simulations and software of this kind in modern NMR spectroscopy.

Methods

SIMPSON [30] and SIMMOL [31] are open source software packages freely available for download from our web site http://nmr.imsb.au.dk. For 3D visualization of SIMMOL output, we recommend the OOGL (Object Oriented Graphics Language) interpreter Geomview [96]. All simulations have been performed on a 1.9 GHz Pentium IV workstation operating under Linux. The procedures and input files described in this paper can be downloaded from our web site [35].

Acknowledgments

This research was supported by grants from Carlsbergfondet, the Danish Research Agency in relation to the Danish Biotechnology Instrument Centre (DABIC), the Danish Natural Science Research Council, and Novo Nordisk Fonden.

References

[1] Ernst RR, Bodenhausen G, Wokaun A (1987) Principles of Nuclear Magnetic Resonanced in One and Two Dimensions. Calendron Press, Oxford
[2] Wütrich K (1986) NMR of Proteins and Nucleic Acids. Wiley, NY
[3] Cavanagh J, Fairbrother WJ, Parmer III AG, Skelton NJ (1996) Protein NMR Spectroscopy. Principles and Practice. Academic Press, San Diego
[4] Haeberlen U, Waugh JS (1968) Phys Rev **175**: 453
[5] Hohwy M, Nielsen NC (1998) J Chem Phys **108**: 3780
[6] Untidt TS, Nielsen NC (2002) Phys Rev E **65**: 021108-1
[7] Sørensen OW, Eich GW, Levitt MH, Bodenhausen G, Ernst RR (1983) Progr Nucl Magn Reson Spectrosc **16**: 163
[8] Wokaun A, Ernst RR (1977) J Chem Phys **67**: 1752
[9] Vega S (1978) J Chem Phys **68**: 5518
[10] Haeberlen U (1976) High-Resolution NMR in Solids. Selective Averaging. Academic Press, NY
[11] Mehring M (1983) High-resolution NMR in solids. Springer, Berlin, Heidelberg, New York, Tokyo
[12] Spiess HW (1978) Rotations of Molecular and Nuclear Spin Relaxation, NMR Basic Principles and Progress. vol 15, Springer, Berlin
[13] Gerstein BC, Dybowski CR (1985) Transient Techniques in NMR of Solids. An Introduction to Theory and Practice. Academic Press, Orlando
[14] Schmidt-Rohr K, Spiess HW (1996) Multidimensional Solid-State NMR and Polymers. Academic Press, London
[15] Raleigh DP, Levitt MH, Griffin RG (1988) Chem Phys Lett **146**: 71
[16] Gullion T, Schaefer J (1989) J Magn Reson **81**: 196
[17] Gregory DM, Mitchell DJ, Stringer JA, Kiihne S, Shiels JC, Callahan J, Mehta MA, Drobny GP (1995) Chem Phys Lett **246**: 654
[18] Bennett AE, Ok JH, Griffin RG, Vega S (1992) J Chem Phys **96**: 8624
[19] Nielsen NC, Bildsøe H, Jakobsen HJ, Levitt MH (1994) J Chem Phys **101**: 1805
[20] Sommer W, Gottwald J, Demco DE, Spiess HW (1995) J Magn Reson A **113**: 131
[21] Lee YK, Kurur ND, Helmle M, Johannessen OG, Nielsen NC, Levitt MH (1995) Chem Phys Lett **242**: 304
[22] Bielecki A, Kolbert AC, Levitt MH (1989) Chem Phys Lett **155**: 341
[23] Hohwy M, Nielsen NC (1997) J Chem Phys **106**: 7571
[24] Larsen FH, Jakobsen HJ, Ellis PD, Nielsen NC (1998) Mol Phys **95**: 1185
[25] Hohwy M, Jakobsen HJ, Edén M, Levitt MH, Nielsen NC (1998) J Chem Phys **108**: 2686
[26] Verel R, Ernst M, Meier BH (2001) J Magn Reson **150**: 81
[27] Vosegaard T, Florian P, Massiot D, Grandinetti PJ (2001) J Chem Phys **114**: 4618
[28] Opella SJ (1997) Nature Struct Biol **4**: 845
[29] Griffin RG (1998) Nature Struct Biol **5**: 508
[30] Bak M, Rasmussen JT, Nielsen NC (2000) J Magn Reson **147**: 296. SIMPSON is open-source software freely available from the web site http://nmr.imsb.au.dk
[31] Bak M, Schultz R, Vosegaard T, Nielsen NC (2002) J Magn Reson **154**: 28. SIMMOL is open-source software freely available from the web site http://nmr.imsb.au.dk

[32] Bak M, Schultz R, Nielsen NC (2001) In: Kiihne SR, de Groot HJM (eds) Perspectives on Solid State NMR in Biology. Kluwer Academic Publishers, Dordrecht, pp 95–109
[33] Welch BB (1995) Practical Programming in Tcl and Tk. Prentice Hall, Englewood cliffs, NJ. The open source Tcl/Tk software can be downloaded via, eg, the Tcl Developers Xhange homepage, http://dev.scriptics.com
[34] James F, Ross M (1975) Comput Phys Commun **10**: 343. Manual available from the web site http://wwwinfo.cern.ch/asdoc/minuit/minmain.html
[35] The SIMPSON and SIMMOL input files for all examples presented in this paper are available for download from the web site http://nmr.imsb.au.dk. SIMPSON and SIMMOL is open source software freely available from the the same web site
[36] Levitt MH (1989) J Magn Reson **82**: 427
[37] Skibsted J, Nielsen NC, Bildsøe H, Jakobsen HJ (1991) J Magn Reson **95**: 88
[38] Charpentier T, Fermon C, Virlet J (1998) J Magn Reson **132**: 181
[39] Levitt MH, Edén M (1998) Mol Phys **95**: 879
[40] Hohwy M, Bildsøe H, Jakobsen HJ, Nielsen NC (1999) J Magn Reson **136**: 6
[41] Zaremba SK (1966) Ann Mat Pura Appl **4–73**: 293
[42] Conroy H (1967) J Chem Phys **47**: 5307
[43] Cheng VB, Suzukawa Jr HH, Wolfsberg M (1973) J Chem Phys **59**: 3992
[44] Wang D, Hanson GR (1995) J Magn Reson A **117**: 1
[45] Alderman DW, Solum MS, Grant DM (1986) J Chem Phys **84**: 3717
[46] Bak M, Nielsen NC (1997) J Magn Reson **125**: 132
[47] Edén M, Levitt MH (1998) J Magn Reson **132**: 220
[48] Kundla E, Samoson A, Lippmaa E (1981) Chem Phys Lett **83**: 229
[49] Charpentier T, Fermon C, Virlet J (1998) J Chem Phys **109**: 3116
[50] Vosegaard T, Skibsted J, Bildsøe H, Jakobsen HJ (1995) J Phys Chem **99**: 10731
[51] Vosegaard T, Skibsted J, Bildsøe H, Jakobsen HJ (1996) J Magn Reson A **122**: 111
[52] Frydman L, Harwood JS (1995) J Am Chem Soc **117**: 5367
[53] Medek A, Harwood JS, Frydman L (1995) J Am Chem Soc **117**: 12779
[54] Marinelli L, Frydman L (1997) Chem Phys Lett **275**: 188
[55] Marinelli L, Medek A, Frydman L (1998) J Magn Reson **132**: 88
[56] Ding S, McDowell CA (1998) J Magn Reson **135**: 1998
[57] Wu G, Rovnyak D, Griffin RG (1996) J Am Chem Soc **118**: 9326
[58] Madhu PK, Goldbourt A, Frydman L, Vega S (1999) Chem Phys Lett **307**: 41
[59] Madhu PK, Goldbourt A, Frydman L, Vega S (2000) J Chem Phys **112**: 2377
[60] Madhu PK, Levitt MH (2002) J Magn Reson **155**: 150
[61] Kentgens APM, Verhagen R (1999) Chem Phys Lett **300**: 435
[62] Vosegaard T, Larsen FH, Jakobsen HJ, Ellis PD, Nielsen NC (1997) J Am Chem Soc **119**: 9055
[63] Larsen FH, Nielsen NC (1999) J Phys Chem A **103**: 10825
[64] Massiot D (1996) J Magn Reson A **122**: 240
[65] Amoureux JP, Pruski M, Lang DP, Fernandez C (1998) J Magn Reson **131**: 170
[66] Goldbourt A, Madhu PK, Kababya S, Vega S (2000) Solid State Nucl Magn Reson **18**: 1
[67] Vega S, Naor Y (1981) J Chem Phys **75**: 75
[68] Nielsen NC, Bildsøe H, Jakobsen HJ (1992) Chem Phys Lett **191**: 205
[69] Dirken PJ, Nachtegaal GH, Kentgens APM (1995) Solid State Nucl Magn Reson **5**: 189
[70] Amoureux J-P, Fernandez C (1998) Solid State Nucl Magn Reson **10**: 211
[71] Oas TG, Griffin RG, Levitt MH (1988) J Chem Phys **89**: 692
[72] Levitt MH, Oas TG, Griffin RG (1988) Isr J Chem **28**: 271
[73] Gan Z, Grandinetti P (2002) Chem Phys Lett **352**: 252
[74] Walls JD, Lim KH, Pines A (2002) J Chem Phys **116**: 79
[75] Nielsen NC, Vosegaard T, Lipton AS, Ellis PD, unpublished results

[76] Vosegaard T, Massiot D, Grandinetti PJ (2000) Chem Phys Lett **326**: 454

[77] Kunath G, Losso P, Steuernagel S, Schneider H, Jäger C (1992) Solid-State Nucl Magn Reson **1**: 261

[78] Opella SJ, Stewart PL, Valentine KG (1987) Quart Rev Biophys **19**: 7

[79] Cross TA, Quine JR (2000) Concepts Magn Reson **12**: 55

[80] Marassi FM, Opella SJ (2000) J Magn Reson **144**: 150

[81] Wang J, Denny J, Tian C, Kim S, Mo Y, Kovacs F, Song Z, Nishimura K, Gan Z, Fu R, Quine JR, Cross TA (2000) J Magn Reson **144**: 162

[82] Wu CH, Ramamoorthy A, Opella SJ (1994) J Magn Reson A **109**: 270

[83] Marassi FM (2001) Biophys J **80**: 994

[84] Vosegaard T, Nielsen NC (2002) J Biomol NMR **22**: 225

[85] Berstein FC, Koetzle TF, Williams GJB, Meier Jr EF, Brice MD, Rodgers JR, Kennard O, Shimanouchi T, Tasumi M (1977) J Mol Biol **112**: 535 Internet address: http://www.rcsb.org/pdb

[86] Vriend G (1990) J Mol Graph **8**: 52

[87] Shaka AJ, Lee CN, Pines A (1988) J Magn Reson **77**: 274

[88] Sattler M, Schleucher J, Griesinger C (1999) Progr NMR Spectrosc **34**: 93

[89] Schwarzinger S, Kroon GJA, Foss TR, Wright PE, Dyson HJ (2000) J Biomol NMR **18**: 43

[90] Baum J, Tycko R, Pines A (1983) J Chem Phys **79**: 4643

[91] Baum J, Tycko R, Pines A (1985) Phys Rev A **32**: 3435

[92] Bendall MR (1995) J Magn Reson A **112**: 126

[93] Kupche E, Freeman R (1996) J Magn Reson A **118**: 299

[94] Tannús A, Garwood M (1996) J Magn Reson A **120**: 133

[95] Cano KE, Smith MA, Shaka AJ (2002) J Magn Reson **155**: 131

[96] Levy S, Munzner T, Philips M et al (1996) Geomview 1.6.1, University of Minesota, Minneapolis. Geomview is open source software freely available from the web site http://www.geomview.org

Residual Dipolar Couplings in ^{31}P MAS Spectra of PPh_3 Substituted Cobalt Complexes

Gábor Szalontai*

University of Veszprém, NMR Laboratory, H-8200 Veszprém, Pf. 158, Hungary

Received June 7, 2002; accepted June 29, 2002
Published online November 7, 2002

Summary. Residual dipolar couplings between ^{31}P–^{59}Co spin pairs were studied in ^{31}P MAS spectra of mono- and dinuclear cobalt-triphenylphosphine complexes. These spectra can provide important informations such as the scalar coupling between the dipolar phosphorus and the quadrupolar cobalt nuclei normally not available from solution phase studies. In case of complementary (NQR or X-ray) data even the relative orientation of the interacting shielding, dipolar, scalar couplings, and electric field gradient tensors or internuclear distances can be determined. Examples are shown both for well resolved and practically unresolved cases, factors which possibly control the spectral resolution are discussed in detail.

Keywords. Solid state NMR; Residual dipolar effects; Cobalt complexes; ^{31}P–^{59}Co pair; One-bond ^{31}P–^{59}Co couplings.

Introduction

Spin–spin interaction between spin-1/2 and quadrupolar nuclei has been the subject of several earlier [1–3] and more recent solid-state NMR studies [4]. The MAS experiment cannot average the dipolar interaction between such spin-pair to zero because the quadrupolar nucleus is not solely quantized by the applied external magnetic field, but also by the anisotropic quadrupolar interaction. The easiest way to look at the effect is to record the MAS spectrum of the spin-1/2 nucleus. Spin pairs, such as ^{13}C–^{14}N [5–8], ^{13}C–^{35,37}Cl [9–12], ^{31}P–^{63,65}Cu [13], ^{119}Sn–^{35,37}Cl [14], ^{31}P–^{35,37}Cl [15], ^{13}C–^{2}H [16, 17], and ^{31}P–^{59}Co [3, 18] have been studied so far.

Our main reason for this work was that the acquisition of MAS spectra allows for the measurements of scalar coupling between the spin-1/2 nucleus ^{31}P and a quadrupolar nucleus, which is not observed normally in liquid phase experiments. In organometallic chemistry the value of the one-bond indirect coupling between phosphorus and a metal is of crucial interest, *e.g.* from the stereochemical point of view. The information one can get from the ^{31}P MAS spectra may even include the

* E-mail: gabor.szalontai@sparc4.mars.vein.hu

orientation of the electric field gradient, EFG tensor, and the magnitude and sign of the quadrupolar coupling constant, and in some cases the asymmetry parameter too. However, in unfavourable conditions, occasionally only good estimates of isotropic shifts and coupling values are available.

We have studied in particular the characteristic of ^{31}P–^{59}Co ($I=7/2$) spin-pair in mono- and dinuclear PPh_3 substituted Co complexes. A practical approach was taken, several examples will be shown for small and larger second-order quadrupolar shifts, for slow and fast cobalt relaxation rates, and for different coordination modes of the cobalt atoms to illustrate the effects and to help the understanding of factors which govern the spectral resolutions.

Results and Discussion

Theory

Using the first-order perturbation treatment [2, 3] first and second-order quadrupolar effects of an S nucleus transferred to MAS spectra of a spin $1/2$ nucleus (I) can be calculated by Eqs. (1)–(5).

The first-order frequency shift, ${}^{1}\Delta\nu_m$ is given by Eq. (1)

$$ {}^{1}\Delta\nu_m = -mJ - mD'(1 - \cos^2\Omega) \tag{1} $$

with Ω the angle between the I, S internuclear vector and the external magnetic field, $\mathbf{B}_o$, D' is the effective dipolar coupling.

The second-order frequency shift, ${}^{2}\Delta\nu_m$ is given by Eq. (2)

$$ {}^{2}\Delta\nu_m = \left(\frac{3D'x}{20\nu_S}\right)\left(\frac{S(S+1) - 3m^2}{S(2S-1)}\right)(3\cos^2\beta^D - 1 + \eta\sin^2\beta^D\cos 2\alpha^D) \tag{2} $$

where m = magnetic quantum number, J = scalar coupling between I and S, ΔJ = anisotropy of scalar coupling, D = dipolar coupling between I and S, its effective value $D' = D - \Delta J/3$, χ = quadrupolar coupling = e^2Qq_{zz}/h, q_{zz} = the largest component of the electric field gradient, Q nuclear quadrupole moment, ν_S = S nucleus *Larmor* frequency, η = asymmetry of the electric field gradient, EFG. For the angles α^D and β^D see the illustration below.

It is known that the MAS experiment averages out the first order effect but only scales down the second-order one. Conditions under which the equation for ${}^{2}\Delta\nu_m$ is valid:

1. First-order perturbation theory can be applied on the S states, *i.e.* the *Zeeman* interaction is much stronger than the quadrupole one [2]:

$$ R_{qz} = \frac{x}{4S(2S-1)\nu_S} \ll 1 \tag{3} $$

2. The $\mathbf{J}$ tensor is axially symmetric and its main axis is aligned with the internuclear vector, r_{IS} (see Scheme 1).

Furthermore, if the EFG tensor $\mathbf{q}$ has axial symmetry too, *i.e.* $\eta = 0$, and is aligned with $\mathbf{r}_{IS}$ then we may define the second-order shift, Δ as follows:

$$ \Delta = \frac{3}{10}\frac{\chi D'}{\nu_S} \qquad \eta = \frac{q_{yy} - q_{xx}}{q_{zz}} \tag{4} $$

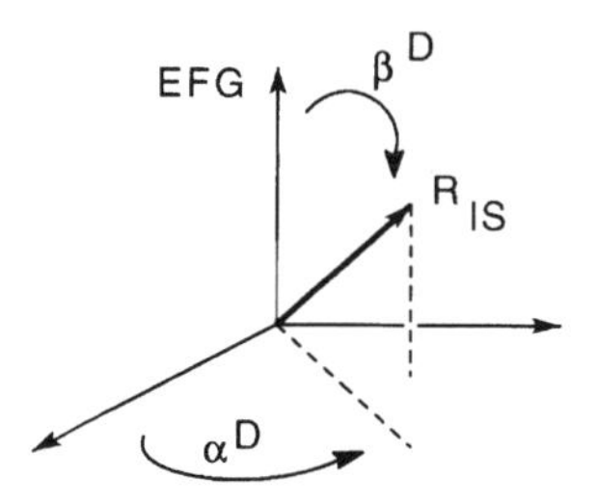

Scheme 1. Co-ordinates of an internuclear distance, R_{IS} in the electric field gradient frame of reference, and its descriptions by the β^D and α^D angles

where as usual $|q_{zz}| \geq |q_{xx}| \geq |q_{yy}|$ (q_x, q_y and q_z are the unit vectors directed along the three axis of the EFG principal axis system).

Using Δ one can calculate positions of all I transitions ($\pm m$) by Eq. (5).

$$\Delta\nu_m + mJ = -\left[\frac{S(S+1) - 3m^2}{S(2S-1)}\right]\Delta \tag{5}$$

Provided the resolution is good enough to see all the $\pm m$ transitions, or at least most of them one can get J_{iso} by inspection of the spectrum, Δ can also be calculated by simple rules. To have a better overview an illustration for the $I(^{31}P)$–$S(^{59}Co)$ case, *i.e.* for a $1/2$–$7/2$ spin pair, is given below (see Scheme 2). It is obvious that the higher the applied magnetic field is, the better is the chance for well resolved transitions.

It is useful to notice that the innermost lines ($m = \pm 1/2$) will always shift in the opposite direction from the outermost lines (for $m = \pm S$ the expression in the square bracket will be -1). General rule for the sense of the second-order shift: if Δ is positive (the "crowding" of peaks occurs to low frequencies) then either χ is

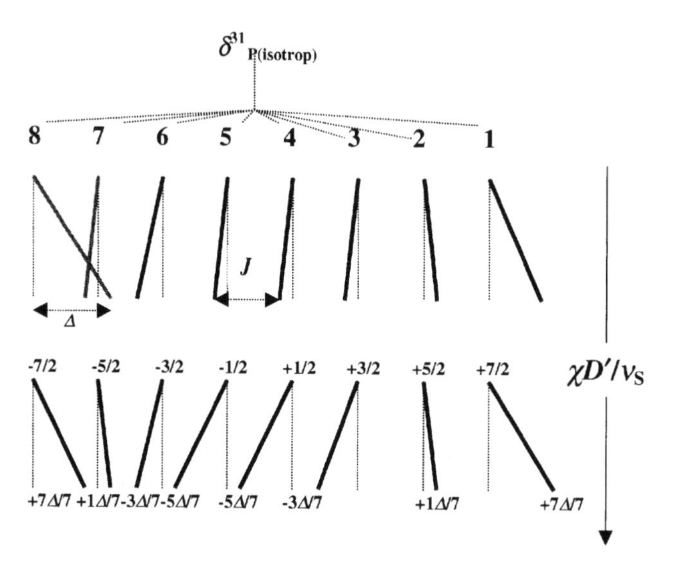

Scheme 2

negative with D and D' of the same sign, or χ is positive while D and D' are of the opposite sign.

Several informations are available from the spectra. First of all the one-bond scalar coupling value between the dipolar (I) and the quadrupolar (S) nucleus $^1J(I,S)$. Furthermore from the second-order shift, Δ, if the quadrupole coupling constant, χ is known (*e.g.* from NQR studies), D' may be derived. If $|\Delta J| \ll |3D|$ (this is very often assumed with good reason) then D is given, leading to a value for r_{IS} (internuclear distance). If r_{IS} is known (*e.g.* from diffraction studies) ΔJ may be derived. Or reversibly, if D' is known (or if D is known and ΔJ can be ignored) the quadrupole coupling, χ may be derived.

Under favourable conditions (first-order perturbation theory applies, J and q tensors have axial symmetry, the dipolar and scalar coupling main axes coincide) scalar and dipolar couplings or the sum of them (note that normally J_{iso} is not available from solution spectra) can be obtained for the spin-pair involved. However, no information can be gained if (a) the relaxation of the quadrupolar nucleus is fast, *i.e.* "self-decoupling" occurs or (b) any solid-state motion averages the mediating dipolar term to zero.

So far we have considered effects on the spacing of transitions of first-order J multiplets. However, even the intensities within the center band or those of within the spinning sidebands can be effected. This may arise from the interplay of the chemical shielding, dipolar, and scalar coupling tensors [19]. Since at room temperature the population should be practically equal for each $2I+1$ transitions the summation of the total spinning sideband intensity corresponding to a specific quantum number should be the same for all other quantum numbers [19]. It means that recording the spectra besides spinning speeds larger than the actual shielding anisotropy should result in equal intensities for all transitions. It also means that deviations from the 1:1:1:1:1:1:1:1 intensity ratio (in case of an isolated cobalt–phosphorus spin pair) are characteristic for the relative orientation of the shielding and dipolar tensors.

Study of Behaviour of $^{31}P(I=1/2)$–$^{59}Co(I=7/2)$ Pairs

Concerning this pair already several reports have been published in literature. *Gobetto et al.* reported on the ^{31}P MAS spectra of bi- and tetranuclear cobalt clusters [3], *Nelson et al.* reported more recently on a series of cobaloximes [18, 20, 21].

Mononuclear Complexes with Trigonal-Bipyramidal Structure, General Formula R–C(O)Co(CO)$_3$P*Ph*$_3$ [22]

For these molecules at 6.33 T or higher we can safely assume the validity of the first-order perturbation theory since even for quadrupole coupling constants of about 200 MHz the ratio between the quadrupole and *Zeeman* interaction, R_{qz} is only about 0.033. The axial symmetry of the dipolar coupling tensor and its co-linearity with the internuclear vector is generally assumed, other necessary conditions such as the axial symmetry of the $\mathbf{J}$ tensor and its co-linearity with the $r_{\mathrm{Co-P}}$ bond are perhaps also met for the one-bond couplings. However, this is not necessarily the case for the symmetry and alignment of the EFG and shielding tensors of the

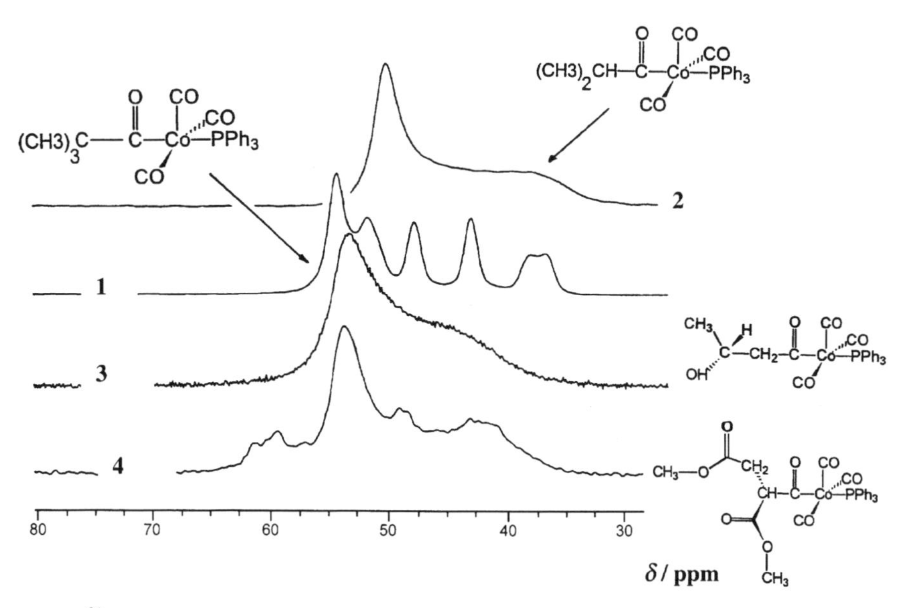

Fig. 1. ^{31}P MAS spectra of **1–4** (centerbands only). Spinning speeds were 5180, 3450, 5150, and 5250 Hz for **1**, **2**, **3**, and **4**, recorded at 109.2 MHz (6.33 T)

Co nucleus. The examples shown above (see Fig. 1) range from the moderately resolved (**2**, $R = -CH(CH_3)_2$) to the completely unresolved (**3**, $R = -CH_2-CH(CH_3)-OH$) cases.

Compound **1** ($R =$ *t*.butyl): The isotropic chemical shift is 50.8 ppm, in the solid phase it is 49.6 ppm. Deconvolution of the experimental spectrum resulted in ten lines instead of the expected eight (see Fig. 2), a possible indication that the assumed axial symmetry of the quadrupole tensor is not complete. Nevertheless, values obtained for the second-order quadrupolar shift (− 342 Hz) and for the one-bond $^{31}P-^{29}Co$ scalar coupling (− 242 Hz) are quite reasonable. The negative sign of Δ comes from the observation that the transitions get closer to each other at high frequencies (peaks 6, 7 and 8 overlap). Anisotropy of the phosphorus shielding, $\Delta\sigma$ is about − 170 ppm (thorough this article we use the *Haeberlen* notation [23], principal components and anisotropies of chemical shielding tensors were calculated from the spinning sideband manifold by the *Herzfeld–Berger* method [24]).

Compound **2** ($R =$ isopropyl): A *J*-coupled fine structure is not resolved, not even in distorted form. The isotropic chemical shift is 49.9 ppm in $CDCl_3$ whereas in the solid phase it is 48.6 ppm. Anisotropy of the phosphorus is about − 175 ppm.

Concerning the possible reasons for the substantial difference between the MAS spectra of **1** and **2**, all other things being very similar, one reasonable explanation might be a difference in the asymmetries of the cobalt EFG tensors due to the different *R* groups. This can lead to fast cobalt relaxation even in solid state what may cause a “self-decoupling” phenomenon [2]. Notice that while in **1** the *R* group has a threefold axis in **2** this is not so.

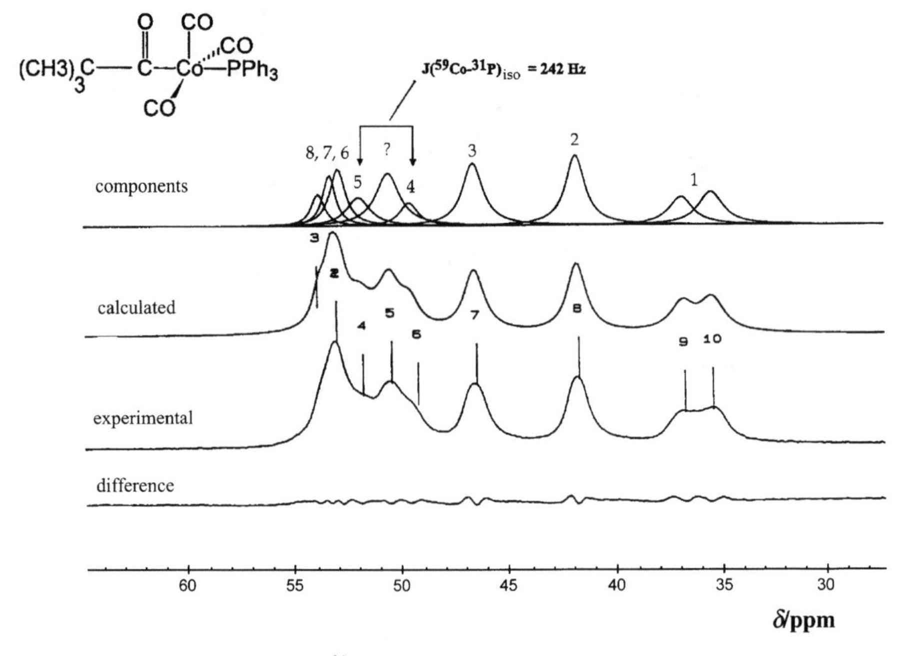

Fig. 2. Deconvolution of ^{31}P MAS spectrum (5180 Hz) of **1** (centerbands only)

Compound **3** (R = –CH_2–$CH(CH_3)$–OH): The isotropic chemical shift is 49.8 ppm, in solid phase it is very close to this. The J-coupling pattern is not resolved here, therefore no estimate for the second-order quadrupolar shift and for the one-bond ^{31}P–^{29}Co scalar coupling could be obtained. Anisotropy of the phosphorus chemical shielding is about – 165 ppm.

Compound **4** (R = –$CH(COOCH_3)$–CH_2–$COOCH_3$ (see Fig. 1)): The isotropic chemical shift is 49.6 ppm, in solid phase it is 49.6 ppm. The observed pattern is not resolved, deconvolution of the experimental spectrum was not possible. It is also possible that more than one crystallographically different molecule is present. Anisotropy of the phosphorus chemical shielding is about – 170 ppm.

Although it is not sufficiently proved, it is likely that small changes of the EFG tensor orientation or its deviation from the axial symmetry are responsible for the substantial changes observed in the spectra of compounds **1**–**4**. At the same time the phosphorus shielding anisotropies and tensor components are only slightly affected.

Compound **5** (CH_2=CH–CH_2–$Co(CO)_2PPh_3$ [25]): The isotropic chemical shift is 67.0 ppm, in solid phase it is 68.8 ppm. This is the least informative case experienced so far. No second-order shift was observed. An indication that either fast cobalt relaxation leads to self-decoupling from the ^{31}P nucleus or solid phase motions are averaging out the residual dipolar couplings. The former is more likely. The observed line width is fairly large (about 605 Hz). Anisotropy of the phosphorus chemical shielding calculated from the spinning sideband manifold is about – 97 ppm, the shielding asymmetry is large (η = 0.87).

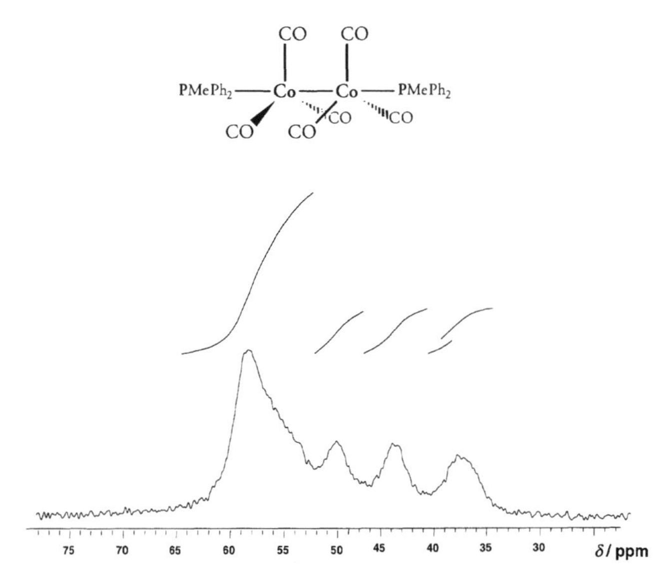

Fig. 3. ^{31}P MAS spectra of **6** (centerbands only). Spinning speed was 8360 Hz, recorded at 121.42 MHz (7.04 T)

Dinuclear Linear Complexes with Co–Co Bond: Ph_2MeP–$Co(CO)_3$–$Co(CO)_3$–P$MePh_2$ (**6**) [26] *vs.* Ph_3P–$Co(CO)_3$–$Co(CO)_3$–PPh_3 (**7**)

The ^{31}P MAS spectrum of **6** is shown in Fig. 3. Due to the heavy overlap of lines which occurs to high frequencies the center transitions are not completely resolved. Nevertheless, a good estimate of J (-395 ± 15 Hz) and the second-order shift, Δ (-440 ± 15 Hz) could be obtained from the spectrum.

Compound **7** (a close analogue to **6**) was thoroughly studied earlier [3], even the value of the quadrupole coupling, χ has been reported (146.8 MHz) and NQR data showed that it has axial symmetry [27]. At 121.4 MHz the phosphorus frequency ν_{Co} is 70.842 MHz therefore R_{qz} is only about 0.024. The P–Co–Co–P direction was thought to be a threefold symmetry axis (X-ray data exist for the (n-butyl)$_3$P analogue [28]. Therefore the co-axiality of tensors was assumed. This is not necessarily the case for **6** where only a C_2 axis can be assumed at best.

Based on arguments used above to explain the observed differences between the spectra of **1** and **2** one would expect different MAS spectra for these compounds too, however this is not the case. In fact the MAS spectra are rather similar. For **6** and **7** the second-order shifts, Δ the isotropic chemical shifts, δ_P and the one-bond scalar couplings, J_{iso}, are -517 Hz, 65.1 ppm, 359 Hz and -440 ± 15 Hz, 52.2 ppm, and 395 ± 15 Hz. This is also a warning that because of the possible interplay of several effects one has to be cautious when trying to interpret spectral changes.

Dinuclear Complexes with Six-Coordinated (Octahedral) Cobalt Atoms

Compound **8**: $Co_2(CO)_2(PPh_3)_2(1,3\text{-}\eta\text{-}S_2CSMe)(\mu\text{-}1,2\text{-}\eta\text{-}SCSMe)$. The X-ray data indicate two nearly identical P atoms with an overall C_s symmetry of the molecule

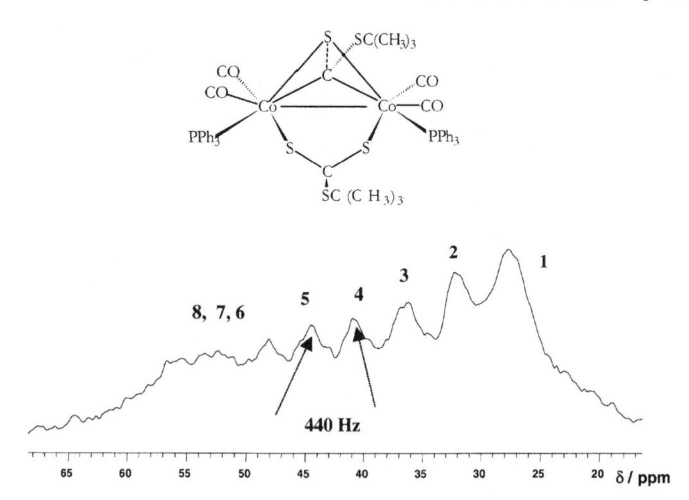

Fig. 4. ^{31}P MAS spectra of **8** (centerbands only). Spinning speed was 5700 Hz, recorded at 121.42 MHz (7.04 T). Nuclear distances from X-ray data: Co–P = 2.232 Å (both), Co–Co = 2.430 Å; the Co–Co–P angles are 154.3 and 158.6 degrees [29]

[29]. The isotropic chemical shift is 38.4 ppm in CD_2Cl_2, whereas in solid phase it is 40.2 ppm. Both Co–P distances are 2.232 Å. The Co–Co–P angles are not identical (154.3 and 158.6 degrees). The ^{31}P MAS spectrum is relatively well resolved (see Fig. 4). Two only slightly different phosphorus environments are suspected, however, only eight lines, though somewhat broadened, are observed. Use of the Co–P distance enable us to calculate D and by assuming that $\Delta J/3$ is much smaller than D, *i.e.* $D' \approx D$, we have an estimate for the quadrupole coupling constant (55.5 MHz). We obtained for J_{iso} and for Δ -440 Hz and -245 Hz.

Compound **9**: $Co_2(CO)_5(PPh_3)_2CHCOOCH_2CH_3$ [30]. X-ray data are not yet available. The isotropic chemical shift is 38.4 ppm in CD_2Cl_2, whereas in solid phase it is 40.2 ppm. The ^{31}P MAS spectrum is relatively well resolved (Fig. 5), two phosphorus environments are present. However, only ten lines are observed. From the distance of the innermost transitions we obtained -449 Hz for **J**, which is rather close to that of compound **8**. Using this value we calculated -75 Hz for the second-order shift, a small value compared to that of **8** ($\Delta = -245$ Hz).

As mentioned already, the interaction between the shielding, σ the dipolar, **D** and scalar coupling, **J** tensors can produce an uneven intensity distribution of the J-coupled multiplet [19]. For compound **9** at low spinning speed (4000 Hz) the central band shows a decrease of multiplet peak intensity in the high-frequency direction. However, by increasing the rotational speed, this is changing gradually, at intermediate frequencies (5800–6000 Hz) the intensities tend to equalize. At a higher rotational frequency (10300 Hz) all spinning sidebands exhibit the reverse trend, *i.e.* the intensities decrease in the low-frequency direction (see Fig. 5 for the changes in the 4000–10300 Hz range). While it is clear that in this molecule the main axis of the dipolar tensor should deviate from that of the shielding tensor, the reason for the complete reversal of the trend is far from being clear.

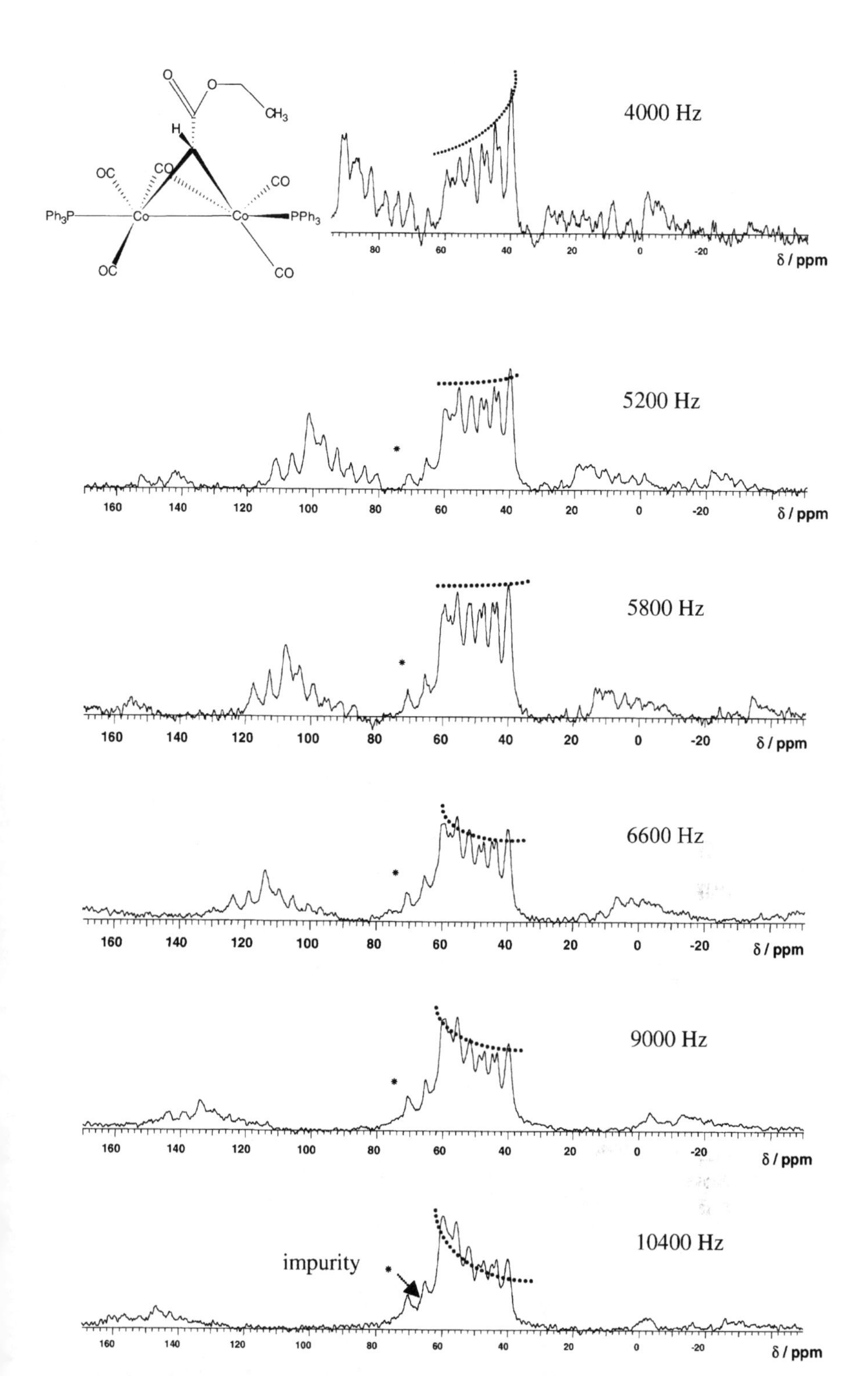

Fig. 5. ^{31}P MAS spectra of **9** recorded at various spinning speeds; 121.42 MHz (7.04 T) [30]

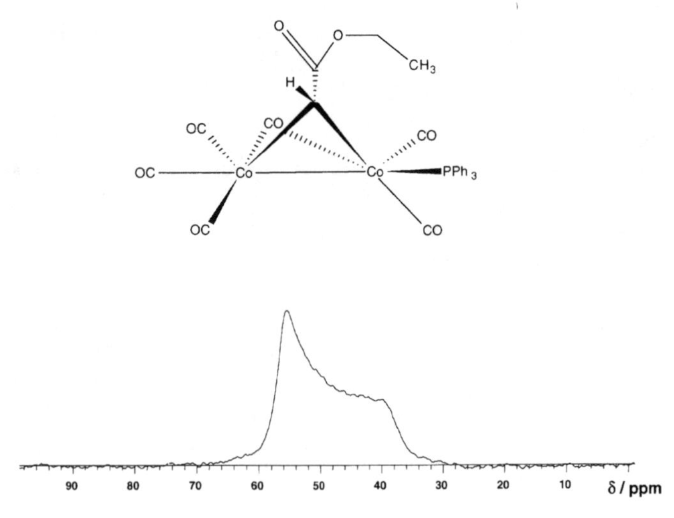

Fig. 6. ^{31}P MAS spectra of (centerbands only). Spinning speed was 9800 Hz, recorded at 121.42 MHz (7.04 T). The Co–Co–P angle was 124.8°

Compound **10**: $Co_2(CO)_6(PPh_3)CHCOOCH_2CH_3$. The X-ray data [30] show a Co–P distance of 2.242 Å, the cobalt atoms are six-coordinated in a strongly distorted octahedral environment. In this case the Co–Co–P angle is only about 124.8 degrees. The isotropic chemical shift is 38.4 ppm in CD_2Cl_2, whereas in the solid phase it is 40.2 ppm. In the ^{31}P MAS spectrum the *J*-coupling is not resolved (see Fig. 6), one phosphorus environment is observed as far as one can judge. The total span of the signal is about 2800 Hz what suggests a $^1J(^{59}Co–^{31}P)$ coupling value of about 400 Hz. This proves that the mediating dipolar field is present, the reason for the loss of coupling information can be the relatively fast cobalt relaxation.

A possible reason for the fast cobalt relaxation can be the strongly distorted octahedral environment of the cobalt atoms (as confirmed by the X-ray data, note the substantial differences in the Co–Co–P angles of **8** and **10**), which likely results in highly asymmetric shielding and EFG tensors.

Experimental

Compounds

$(CH_3)_3C–C(O)Co(CO)_3PPh_3$ (**1**, $C_{26}H_{24}O_4CoP$) [21], $(CH_3)_2CH–C(O)Co(CO)_3PPh_3$ (**2**, $C_{25}H_{22}O_4CoP$) [21], $CH_3–CH(OH)–CH_2–C(O)Co(CO)_3PPh_3$ (**3**, $C_{25}H_{22}O_5CoP$) [21], $CH_3O–CO–CH_2–CH(COOCH_3)–C(O)Co(CO)_3PPh_3$ (**4**, $C_{28}H_{24}O_4CoP$) [21], $CH_2{=}CH–CH_2–Co(CO)_2PPh_3$ (**5**, $C_{23}H_{20}O_2Co_1P_1$) [25], $Ph_2MeP–Co(CO)_3–Co(CO)_3–PMePh_2$ (**6**, $C_{32}H_{26}O_6Co_2P_2$) [26], $Ph_3P–Co(CO)_3–Co(CO)_3–PPh_3$ (**7**, $C_{42}H_{30}O_6Co_2P_2$) [3], $Co_2(CO)_2(PPh_3)_2(1,3\text{-}\eta\text{-}S_2CSMe)(\mu\text{-}1,2\text{-}\eta\text{-}SCSMe)$ (**8**, $C_{50}H_{48}O_4S_5Co_2P_2$) [29], $Co_2(CO)_5(PPh_3)_2CHCOOCH_2CH_3$ (**9**, $C_{45}H_{36}O_7Co_2P_2$) [30], $Co_2(CO)_5(PPh_3)_2CHCOOCH_2CH_3$ (**10**, $C_{28}H_{21}O_8Co_2P$) [30].

Spectroscopy

Most of the spectra were recorded on a Varian UNITY 300 spectrometer using a Doty XC5 room temperature probe under the conditions of high-power proton decoupling and magic-angle spinning.

The ^{31}P 90° pulse duration was about 3.5 μsec, spectral width 50000 Hz, acquisition time 0.05 sec, number of transients 128–512, recycle delay 20–60 sec. Depending on the line widths obtained 10–50 Hz line broadening function was applied. The rotation rates were varied between 3000 and 11000 Hz. Centerbands were located by changing the sample rotation rate. Phosphorus chemical shifts were obtained by the substitution method and are quoted relative to the 85% H_3PO_4. The proper MAS conditions were checked with crystalline PPh_3 ($\delta_P = -6$ ppm) put in an insert of about 35 mm^3. For this sample resolutions better than 30 Hz were obtained using 5 mm Si_3N_4 Doty rotors. The 109.38 MHz ^{31}P MAS spectra were obtained on a JEOL GX 270/89 spectrometer using 6 mm o. d. zirkonia rotors. The accuracy of the J values obtained clearly depend on the spectral resolution and are indicated in the text where appropriate, consequently the accuracy of the calculated second-order shift values is not better than $\pm$ 15 Hz.

Acknowledgement

The author wishes to acknowledge the assistance and advices of Professors *S. Aime* and *R. Gobetto* (Torino) in the NMR determination of compounds **1–4**, and Professors *L. Markó* and *F. Ungváry* and *A. Sisak*, *S. Vastag*, and *I. Kovács*, and PhD student *R. Tuba* (Veszprém) for kindly providing X-ray results and samples **5–10**. The Hungarian Scientific Research Fund (OTKA) is gratefully acknowledged for financial support (project number T34355).

References

[1] Menger EM, Veeman WS (1982) J Magn Reson **46**: 257
[2] Harris RK, Olivieri A (1992) Progress in NMR Spectroscopy **24**: 435
[3] Gobetto R, Harris RK, Apperley D (1992) J Magn Reson **96**: 119
[4] Freude D, Haase J (1999) Quadrupole Effects in Solid-State Nuclear Magnetic Resonance. In: NMR Basic Principles and Progress **29**: Springer, Berlin
[5] Davies NA, Harris RK, Olivieri AC (1996) Mol Phys **87**: 669
[6] Suits BH, Sepa J, White D (1996) J Magn Reson **120**: 88
[7] McDowell CA (1996) In: Encyclopedia of Nuclear Magnetic Resonance. Grant DM, Harris RK (eds) 5: Wiley, Chichester, UK p 2901
[8] Hexem JG, Frey MH, Opella SJ (1982) J Chem Phys **77**: 3487
[9] Alarcón SH, Olivieri AC, Crass SA, Harris RK (1994) Angew Chem **106**: 1708
[10] Alarcón SH, Olivieri AC, Crass SA, Harris RK, Zuriaga MJ, Monti GA (1995) J Magn Reson A **116**: 244
[11] Nagasaka B, Takeda S, Nakamura N (1994) Chem Phys Lett **222**: 486
[12] Cravero RM, González-Sierra M, Fernández C, Olivieri AC (1993) J Chem Soc Chem Commun 1253
[13] Olivieri AC (1992) J Am Chem Soc **114**: 5758
[14] Komorovski RA, Parker RG, Mazany AM, Early TA (1987) J Magn Reson **73**: 389
[15] Thomas B, Paasch S, Steuernagel S, Eichele K (2001) Solid State Nuclear Magnetic Resonance **20**: 108–117
[16] Jonsen P, Tanner SF, Haines AH (1989) Chem Phys Lett **164**: 325
[17] Swanson SD, Ganapathy S, Bryant RG (1987) J Magn Reson **73**: 239
[18] Nelson JH (2002) Concepts in Magnetic Resonance **14**: 19 (see further references therein)
[19] Chu PJ, Lunsford JH, Zalewski DJ (1990) J Magn Reson **87**: 68
[20] Schurko RW, Wasylishen RE, Nelson JH (1996) J Phys Chem **100**: 8057
[21] Schurko RW, Wasylishen RE, Moore SJ, Marzilli RG, Nelson JH (1999) Can J Chem **77**: 1973
[22] Kovács I, Szalontai G, Ungváry F (1996) J Organomet Chem **524**: 115
[23] Haeberlen U (1976) in High Resolution NMR in Solids (Suppl 1) Academic Press, New-York
[24] Herzfeld J, Berger AE (1980) J Chem Phys **73**: 6021

[25] Heck RF, Breslow DS (1961) J Am Chem Soc **83**: 1097
[26] Hieber W, Freyer W (1958) Chem Ber **91**: 1230
[27] Ogino K, Brown TL (1971) Inorg Chem **10**: 517
[28] Ibers JA (1968) J Organomet Chem **14**: 423
[29] Gervasio G, Vastag S, Szalontai G, Markó L (1997) J Organomet Chem **533**: 187
[30] Ungváry F, Tuba R (unpublished results)

Potential and Limitations of 2D ^1H–^1H Spin-Exchange CRAMPS Experiments to Characterize Structures of Organic Solids

Jiri Brus[1,*], **Hana Petříčková**[2], and **Jiri Dybal**[1]

[1] Institute of Macromolecular Chemistry, Academy of Sciences of the Czech Republic, CZ-162 06 Prague 6, Czech Republic
[2] Institute of Chemical Technology, CZ-Prague, 166 28 Prague 6, Czech Republic

Received May 28, 2002; accepted (revised) July 1, 2002
Published online November 7, 2002

Summary. A brief overview of our recent results concerning the application of 2D CRAMPS experiments to investigate a wide range of materials is presented. The abilities of the 2D ^1H–^1H spin-exchange technique to characterize the structure of organic solids as well as the limitations resulting from segmental mobility and from undesired coherence transfer are discussed. Basic principles of ^1H NMR line-narrowing and procedures for analysis of the spin-exchange process are introduced. We focused to the qualitative and quantitative analysis of complex spin-exchange process leading to the determination of domain sizes and morphology in heterogeneous multicomponent systems as well as the characterization of clustering of surface hydroxyl groups in polysiloxane networks. Particular attention is devoted to the determination of the ^1H–^1H interatomic distances in the presence of local molecular motion. Finally we discuss limitations of the ^{13}C–^{13}C correlation mediated by ^1H–^1H spin exchange to obtain structural constraints. The application of *Lee-Goldburg* cross-polarization to suppress undesired coherence transfer is proposed.

Keywords. CRAMPS; 2D solid-state NMR; *Lee-Goldburg*; Miscibility of polymers; Clustering of surface silanols; Molecular dynamics.

Introduction

The importance of NMR follows from its unique selectivity differentiating various chemically distinct sites on the basis of their chemical shifts. In solution-state NMR the protons are most important due to a nearly 100% natural abundance and the highest gyromagnetic ratio γ. From this follows the best sensitivity of

* Corresponding author. E-mail: brus@imc.cas.cz

all naturally occurring nuclei. On the other hand, just this combination of properties dramatically complicates recording of ^{1}H NMR spectra in the solid state. Missing isotropic tumbling leads to a severe broadening of the NMR signals as a result of non-averaged anisotropic interactions. The dominant anisotropic interaction in solid-state ^{1}H NMR is dipolar coupling hindering resolution of chemically different sites. That is why one broad signal with a line-width of several tens of kHz is usually observed. The lack of ^{1}H NMR spectrum resolution does not mean the absence of structure information in the obtained spectra, it rather reflects its overcrowding in such an extent that we are not able to read out and understand it. Due to this fact various techniques have been proposed to increase the spectral resolution of ^{1}H NMR spectra because they provide valuable indications about the local chemical environment and a wide range of structural information.

In recent years we have investigated several systems by solid-state ^{1}H NMR spectroscopy. In the first part of this contribution we briefly summarize the basic concept of techniques leading to the averaging of anisotropic nuclear interactions and we introduce description and analysis of the spin-exchange process. In the second part we present our recent results concerning applications of two-dimensional (2D) ^{1}H spin-exchange experiments applied to the characterization of the structure and geometry of a wide range of materials. At first qualitative and quantitative data providing information about miscibility, morphology and domain sizes are discussed. Further, the investigation of clustering of surface hydroxyl groups including silanols and adsorbed water molecules in polysiloxane networks is introduced. Particular attention is paid to the precise measurement of ^{1}H–^{1}H interatomic distances in crystalline organic solids in the presence of local molecular motion. Finally we discuss the application of ^{13}C–^{13}C correlation mediated by ^{1}H–^{1}H spin exchange to obtain precise data leading to the extraction of ^{1}H–^{1}H spatial separation. The advantage of *Lee-Goldburg* irradiation to suppress undesired coherence transfer during cross-polarization steps is introduced.

Methods and Principles

Magic Angle Spinning (MAS)

One possible way how to compensate missing molecular motion is mechanical uniaxial rotation. It is known that anisotropic interactions such as dipolar coupling between a pair of nuclei or chemical shift anisotropy (CSA) have an orientation dependence that can be described by the second rank tensor [1, 2]. In such case physical rotation of the sample around the axis, which is inclined at the angle 54.7° (magic angle) with respect to static magnetic field leads to an averaging of anisotropy broadening to zero [3, 4]. If anisotropic interaction is refocused at the end of each rotor period (*i.e.* CSA) the originally broad static NMR signal is easily broken up into a sharp central signal reflecting isotropic chemical shift and series of spinning sidebands separated by the roation frequency ν_r. The line-widths of these signals are independent of the spinning speed (see Fig. 1A).

However, in the case of a multibody strongly dipolar-coupled spin-system the situation is quite different. In contrast to the previous case of CSA the line-widths of all ^{1}H signals are ν_r frequency dependent and the central signal and spinning

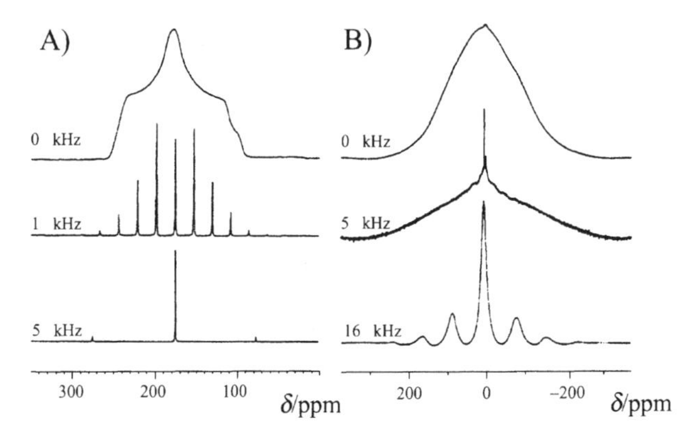

Fig. 1. ^{13}C NMR (A) and ^{1}H NMR (B) spectra of ^{13}C selectively labeled (C=O) glycine measured at various MAS spinning frequencies

side bands are still relatively broad. Even at moderately fast spinning speeds the ^{1}H NMR signals remain much larger than those observed in ^{13}C MAS NMR spectra. From Fig. 1B it is clear that a spinning speed of *ca.* 15 kHz, which is achieved with a standard 4-mm probe-head, is still not sufficient to provide reasonable good resolution. Recent development of new 2.5-mm probe-heads has made it possible to routinely achieve much greater rotation frequencies up to 35 kHz and even a rotation frequency of 50 kHz has been reported [5]. Such fast MAS leads to relatively well resolved spectra. However, the resolution of ^{1}H MAS NMR spectra still is not comparable with the resolution of spectra obtained for liquid samples and in addition ultra-high spinning speed induced substantial friction heating of the sample [6–8]. Temperature increase within the sample due to fast rotation may make-up to about 60 K, and temperature gradients within the samples (up to 17 K) may cause an additional broadening of signals [9]. That is why alternative procedures of narrowing ^{1}H signals in the solid state should be used in particular cases.

Multipulse ^{1}H Homodecoupling (Combined Rotation and Multiple-Pulse Spectroscopy)

The first classical multiple-pulse homodecoupling technique (WHH-4) was designed in 1968 [10]. It consists of many cycles, each made up of two solid-echo pulse pairs: that is, four 90° pulses are separated by windows of duration τ or 2τ (see Fig. 2A). As we would like to avoid a quantum mechanical treatment, we use only highly simplistic and intuitive description of the techniques. For more detailed analysis see *e.g.* [11–13]. In the analysis of these techniques, the duration of the pulses is assumed to be negligible. After a cycle of 6τ (for sufficiently short pulse-spacing τ) the spins effectively evolve under the average Hamiltonian reflecting only chemical shift and resonance offset. Stroboscopic detection at these times ($t = 6\tau$, in Fig. 2 asterisks indicate acquisition of one data point) therefore produces a time signal that is modulated only by chemical shift, resonance offset and heteronuclear dipole couplings. During the multiple-pulse sequence the precession no longer occurs around the direction of the static magnetic field; instead, the

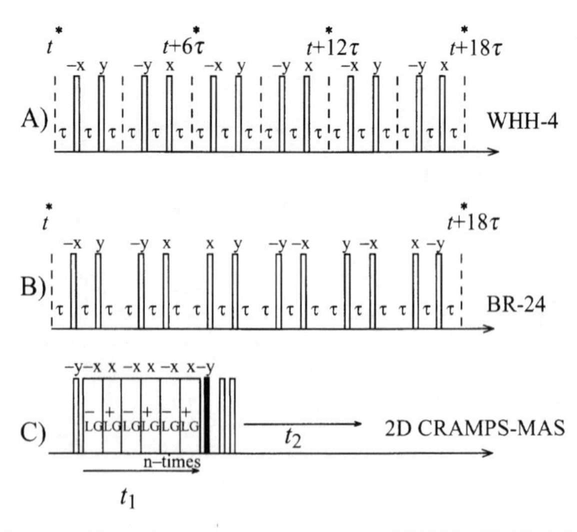

Fig. 2. Pulse schemes of homodecoupling sequences: (A) WHH4; (B) BR-24 (one half of the sequence) and (C) 2D CRAMPS-MAS experiment exploiting *Lee-Goldburg* decoupling during t_1 detection (the black block corresponds to the 54° pulse)

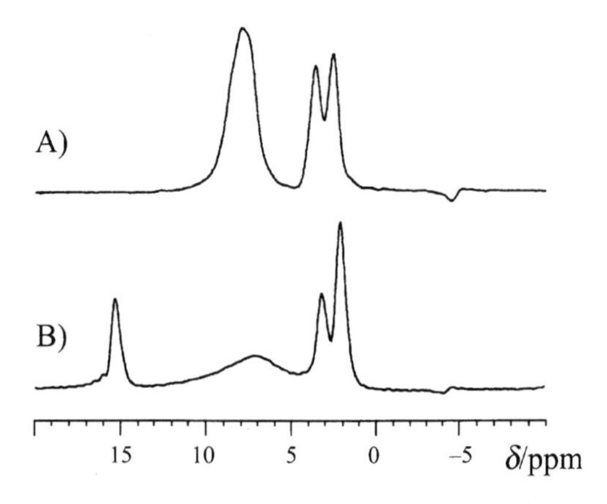

Fig. 3. [1]H CRAMPS spectra of glycine (A) and aspartic acid (B) measured by the BR-24 pulse sequence ($\pi/2$ [1]H pulse – 1.8 μs, small and large window 3.8 and 1.0 μs, MAS 2 kHz)

first-order average Hamiltonian causes a precession with an effective frequency $\omega_{\text{eff}}^{j} = \omega^{j}/\sqrt{3}$ around the effective-field unit vector $(1/\sqrt{3})(1,1,1)$. The ratio $\omega_{\text{eff}}^{j}/\omega^{j}$ is generally known as the frequency scaling factor. The applied train of *rf* pulses causes the rotation in spin space [10, 11] and dipolar broadening is suppressed. Well resolved ^{1}H MAS NMR spectra are achieved by simultaneous application of MAS refocusing chemical shift anisotropy [14–16] (see Fig. 3). This combination is termed as "combined rotation and multiple-pulse spectroscopy" (CRAMPS).

Several preconditions and approximations in the derivation of action of these pulse sequences indicate the special requirements for the use of multipulse techniques: pulses with accurate flip angles and relative phases are needed. Also the pulse length and the spacing τ must be as small as possible. To reduce effects of pulse imperfection and higher-order terms of dipolar couplings, other pulse sequences based on WHH-4 have been developed. Among the various techniques, the MREV-8 sequence (*Mansfield-Rhim-Elleman-Vaughan*; consisting of two phase-cycled WHH-4) [17] seems to be highly robust. Further improvement provides technique BR-24 developed by *Burum* and *Rhim* [18] consisting of three partly nested MREV-8 cycles, which averages out the homonuclear interaction up to the third order (see Fig. 2B). Therefore this sequence provides best results in a well tuned spectrometer.

As mentioned above, sufficiently fast MAS considerably reduces dipolar broadening. From this fact one would expect that a multipulse sequence could be performed more easily at a high spinning speed leading to much more resolved spectra. However, both averaging techniques mutually interfere [19], which causes a rather dramatic loss of spectral resolution when a multipulse sequence is simply applied at high speed MAS conditions [20]. In fact, a spinning frequency less than 3 kHz is usually used. Under this condition (quasi-static limit) the sample is considered to be static during each cycle of the multipulse sequence. The mutual interference can be reduced by synchronization of the multiple-pulse sequence with MAS (variants of WHH-4 sequence were applied at moderately fast MAS, 10–15 kHz) [20–22]. Under such conditions the original philosophy of CRAMPS experiments is shifted. At fast MAS multipulse sequence may not completely remove dipolar broadening as required in quasi-static limit; rather, it is sufficient if the given sequence reduces dipolar couplings to such an extent that fast MAS can remove the residual contribution. Such experimental techniques are known as "multiple-pulse assisted MAS".

An alternative approach leading to averaging of homonuclear dipolar couplings is based on the *Lee-Goldburg* experiment [23], in which the offset of ^{1}H *rf* irradiation is set equal to $\omega_1/\sqrt{2}$, where ω_1 is the nutation frequency of the ^{1}H pulse ($|\omega_1| = |\gamma B_1|$). Using the vector model, the ^{1}H spins rotate under this irradiation around an effective field inclined at the angle 54.7° with respect to the static magnetic field. A significant enhancement of this technique was achieved by the frequency-switched modification of the *Lee-Goldburg* experiment (FSLG) [24, 25]. Instead of continuous irradiation, a series of 2π ^{1}H pulses with an offset switched between two LG conditions $\mp\omega_1/\sqrt{2}$ accompanied by a phase shift of π is applied. This technique works well for spinning speeds ranging from 10 to 16 kHz [26]. An alternative interpretation of the *Lee-Goldburg* technique was recently presented by *Vinogradov* and co-workers [28]. In this experiment (phase-modulated *Lee-Goldburg* – PMLG) only the phase of a series of adjacent pulses is changed. The frequency of the B_1 field remains constant. It was shown that the zero-order term of the average Hamiltonian vanishes when the modulation of the pulse phase $\phi(t) = \omega_{PMLG} t$ statisfies the condition: $|\omega_{PMLG}| = \omega_1/\sqrt{2}$ and the duration of the LG irradiation unit corresponds to a 2π rotation of the proton magnetization about the effective field, *i.e.* $t_{LG} = \sqrt{(2/3)}(2\pi/\omega_1)$. From this it is evident that the angle through which *rf* precesses in one LG unit is given by $\alpha_{LG} = |\omega_{PMLG}| t_{LG} = 207.8°$. As symmetrization is required to ensure the removal of odd-order terms in the dipolar Hamiltonian, the sign of the phase modulation has to be negated between alternate LG units. Hence, during the first part of the PMLG sequence, the *rf* field precesses from 0° to 208° and then to 180°, after a 180° flip, during the second half of the PMLG unit. This constitutes a unit of PMLG which is executed by a series of short pulses with well-defined phases, α_i, for the *i*-th pulse. The duration of each pulse has been chosen to be *ca.* 1 μs and *rf* field strength 82 kHz. These originally developed pulse sequences applying both concepts of *Lee-Goldburg* irradiation (FSLG and PMLG) have no windows for direct signal detection; that is why ^{1}H NMR spectra are obtained using an indirect detection scheme (see Fig. 2C). The resolution of the obtained spectra (t_1 projection of the 2D experiments) is very promising. For instance, the halfwidth of ^{1}H NMR signals of malonic acid is 0.3 ppm [28]. Recently, a wide range of two-dimensional (2D) and three-dimensional (3D) experiments has been designed on the basis of these homodecoupling schemes [26–30]. Quite recently, successful insertion of detection windows in the PMLG schemes has been reported [31]. It has to be noted that the first attempt in this direction was made by *Levitt et al.* [32] by inserting windows in the FSLG sequence. By this way high-resolution ^{1}H NMR spectra can be measured in a 1D fashion. Application of windowed PMLG simplifies ^{1}H–^{1}H correlation experiments for signal assignment and distance measurements and enables an inverse detection to enhance the sensitivity to experiments.

^{1}H–^{1}H Spin Exchange (Spin Diffusion)

Although the strong spin interaction (given by the combination of nearly 100% natural abundance of ^{1}H nuclei and the highest gyromagnetic ratio γ) highly complicates the recording of ^{1}H NMR spectra for the solid state, it offers interesting structural information. For instance, it can be provided by the spatial transfer of z magnetization between dipolar-coupled spins. Such a transfer is termed "spin exchange" or "spin diffusion" [33] and it is most efficient between protons. The quantum mechanical treatment of spin exchange between two spins revealed its oscillatory behavior. However, for systems of many spins the complicated network of couplings cancels all oscillations thus leading effectively to diffusive behavior [2].

It has to be stressed that ^{1}H spin diffusion in the solid-state is a coherent, fully reversible process [34] in contrast to relayed ^{1}H polarization transfer based on the multistep nuclear *Overhauser* effect (NOE) observed in certain solution-state NMR experiments, which is also termed spin diffusion. The latter process is an incoherent cross-relaxation induced by the stochastic modulation of local fields by molecular motion. Due to its stochastic time dependence this process cannot be refocused.

Spin-diffusion experiments are typical exchange experiments, consisting of an evolution or selection period, a mixing time t_m, and a detection period. The time dependence of the spin-diffusion process contains information on the domain size in heterogeneous materials: in systems with small domains, the magnetization equilibrates faster than in a system consisting of large particles. Generally, spin diffusion typically probes the smallest distances since the equilibrium occurs *via* the shortest path. For spin diffusion to occur, a spatially inhomogeneous z magnetization distribution has to be generated. The magnetization of one component is selected while the magnetization of the second component is suppressed. During a

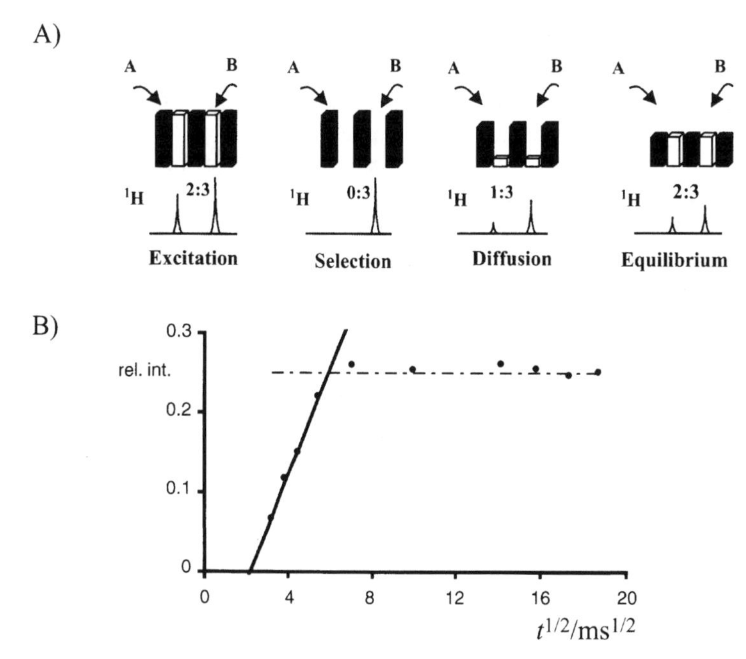

Fig. 4. (A) Basic scheme of the spin-diffusion experiment: Magnetization of the A component is selected during the selection period to generate the magnetization gradient. During mixing time t_m spin-diffusion occurs, which is detected by an increase of the signal intensity of the B component. (B) Spin-exchange built-up data

mixing period, magnetization of the selected component is transferred by double and zero-quantum transitions to neighboring spins. At short mixing times, ^{1}H polarization is transferred between the nearest spins, while at longer times, relayed polarization transfer to further spins occurs at a rate proportional to $1/r_{ij}^3$. Then, the distribution of magnetization of one component is monitored in the NMR spectrum (see Fig. 4). Thus, ^{1}H–^{1}H spin-exchange experiments can be used to measure short-range distances providing constraints for structure determination and signal assignment using detection of spin-exchange process at very short mixing times, while relayed ^{1}H–^{1}H spin exchange (spin diffusion) makes it possible to study long-range ordered structures on a 0.5–200 nm scale.

Analysis of Spin-Exchange Built Up-Curves

In order to obtain information about the domain size from spin-exchange data, a simulation of the time-dependence of polarization exchange has to be performed. In general, the rate of magnetization exchange $P(t)$ between two nuclei j and k is given by Eq. (1)

$$P(t) = \frac{1}{2}\pi g_0^{jk}(\omega_j - \omega_k)\omega_D^2 t \tag{1}$$

where the term $g_0^{jk}(\omega_j - \omega_k)$ is proportional to the overlap between the j-th and k-th ^{1}H NMR signals, while ω_D measures the dipolar coupling between the two nuclei

[35]. In a strongly dipolar-coupled multibody spin-system, however, the phenomenological description based on *Fick*'s second law is usually used. Up to now, several approaches and procedures to extract the desired information have been proposed: a general initial-rate approximation, a rigorous solution for lamellar structures, related approximation for more complicated disordered morphologies, and a versatile lattice-calculation approach for arbitrary distributions of diffusivity values and initial z magnetization [2, 36–39]. In the simplest initial-rate approximation, analyzing the straight-line part of the spin-exchange built-up curve, the displacement of polarization between two neighbouring spins can be described by relation (2)

$$r = \left(\frac{4}{3} D t_m\right)^{1/2} \tag{2}$$

proposed by *Van der Hart et al.* [40], which is a special case of a generally derived spin-diffusion equation for two-component systems [2, 41, 42]:

$$d_\mathrm{A} = 2\frac{\varepsilon}{f_\mathrm{B}}\left(\frac{1}{\pi} D t_m^s\right)^{1/2} \tag{3}$$

d_A is the size of A component, f_B the volume fraction of B component, t_m^s the time of the straight line intersection with the $I = 100\%$ (*i.e.* $I = I_\mathrm{B}$ $(t_m \to \infty)$) and ε is the dimensionality of the spin-exchange process (*i.e.* the number of orthogonal directions relevant for the magnetization transfer). From the dimensionality ε, one can estimate the morphology of the studied system (lamellar – $\varepsilon = 1$; cylindrical – $\varepsilon = 2$; spherical – $\varepsilon = 3$). However, a recent study [37] has indicated that the determination of the morphology of an unknown system is very rough, because all models are based on assumptions which are not quite realistic (*e.g.*, regular repetition of the domains, the same shape of the domains, and Gaussian distribution of their size, etc.). The crucial parameter for accurate analysis is the spin-diffusion (spin-exchange) coefficient D reflecting the strength of ^{1}H–^{1}H dipolar interactions within each component. For the analysis of clearly motionally heterogeneous two-component systems, the knowledge of the diffusivity of both components is required. If the diffusivities of the components differ ($D_\mathrm{A} \neq D_\mathrm{B}$), an effective spin diffusion D_eff coefficient has to be used.

$$\sqrt{D_\mathrm{eff}} = \frac{2\sqrt{D_\mathrm{A} D_\mathrm{B}}}{\sqrt{D_\mathrm{A}} + \sqrt{D_\mathrm{B}}} \tag{4}$$

The exact diffusivity D was determined from spin-diffusion built-up curves only for a well-defined material with known morphology and domain size, such as the diblock copolymer *PS-PMMA* (polystyrene-polymethylmethacrylate) the structure of which as been investigated by small-angle X-ray scattering and transmission electron microscopy [36]. The obtained value $D = 0.8 \pm 0.2\,\mathrm{nm^2\,ms^{-1}}$ is generally used for highly rigid organic solids. However, for systems with unknown geometry other approaches have to be used. In general, the diffusivity is expressed in terms of local dipolar fields proposed by *Cheung* [43, 44]. For instance, the following Eqs. (5) and (6), relating D with the ^{1}H line-width at half intensity

$\Delta\nu_{1/2}$ were proposed for the Gaussian line-shape reflecting the rigid component [38, 45].

$$D_{\text{rig}} = \frac{1}{12}\sqrt{\frac{\pi}{2\ln 2}}\langle r^2\rangle\Delta\nu_{1/2} \tag{5}$$

$$D_{\text{rig}} = \Delta\nu_{1/2}\frac{\langle r^2\rangle}{3} \tag{6}$$

Here $\langle r^2\rangle$ is the square average of the ^{1}H–^{1}H internuclear distance, which is assumed to be ranging in organic solids from 0.2 to 0.25 nm. For mobile components characterized by Lorentzian ^{1}H NMR line-shape, the diffusivity is expressed by Eq. (7), where α is the cutoff parameter.

$$D_{\text{mob}} = \frac{1}{6}\langle r^2\rangle[\alpha\Delta\nu_{1/2}]^{1/2} \tag{7}$$

Alternative expressions relating diffusivity to T_2 relaxation have been reported by *Mellinger et al.* [41]:

$$D_{\text{mob}} = 8.2\times10^{-6}T_2^{-1} + 0.007 \tag{8}$$

$$D_{\text{mob}} = 4.4\times10^{-5}T_2^{-1} + 0.26 \tag{9}$$

Relations (8) and (9) are supposed to be valid for regions corresponding to $0 < T_2^{-1} < 1000$ and $1000 < T_2^{-1} < 3500$ Hz.

For mobile amorphous poly(ethylene oxide) (*PEO*) with the ^{1}H NMR linewidth of *ca.* 0.6–1.7 kHz and amorphous polyethylene (*PE*) with the linewidth of *ca.* 1.8 kHz, diffusivities in the range of 0.09–0.15 nm^2 ms^{-1} have been reported [36, 46–48]. Lower values ranging from *ca.* 0.03 to 0.08 nm^2 ms^{-1} have been determined for mobile *PEO* and polyisoprene by *Mellinger et al.* [41]. *Spiegel et al.* [49] have found a diffusivity of 0.05 nm^2 ms^{-1} for polyisoprene. For soft polyurea segments [50] and highly mobile aliphatic side-chains [51] spin-diffusion coefficients have been determined to be 0.04 and 0.05 nm^2 ms^{-1}. For a rubbery poly(epichlorohydrin)/poly(vinyl acetate) blend [52], even a value of 0.01 nm^2 ms^{-1} has been reported. Mobile poly(dimethylsiloxane) chains cross-linking polyimide chains have been characterized by $D = 0.09$ nm^2 ms^{-1} [53]. A diffusivity of 0.4 nm^2 ms^{-1} has been determined [51] for the aromatic main-chain protons of hard segments of poly(1,4-phenyleneterephthalimide) and poly(1,4-phenylenepyromellitimide), as well as for an aromatic poly(ester-urethane) elastomer [43, 54]. For crystalline domains of *PEO*, spin-diffusion coefficients [46] are assumed to be 0.29–0.32 nm^2 ms^{-1}. Much higher diffusivities have been reported for rigid crystalline solids and densely packed polymers; the diffusivity of *PE* [46, 47] is 0.7–0.8 nm^2 ms^{-1} and of crystalline alanine [55] it is 0.6–0.8 nm^2 ms^{-1}.

Results and Discussion

Due to the relatively low resolution of ^{1}H NMR spectra of solids, one-dimensional (1D) spin-diffusion experiments have been predominantly used up to the recent past to study the homogeneity of various mixtures and the miscibility of their components [36, 40, 56]. The application of such 1D experiments only for the investigation of relatively simple systems follows also from the necessity to create an initial magnetization gradient to provoke spin diffusion. This selection period is often based on T_2 or T_1 relaxation [57, 58] or multiple-pulse dipolar filters [50, 51, 58, 59], which discriminate components substantially differing in mobility. That is why spin diffusion has been predominantly studied only between two dynamically different components, although techniques based on chemical shift filtering [36, 58, 60] and selective saturation transfer [61, 62] have been also applied. A significant increase in resolution and generalization of spin-exchange experiments has been provided by two-dimensional (2D) technique proposed by *Caravatti et al.* [63] making it possible to observe polarization transfer between all sites resolved in a 2D ^1H–^1H CRAMPS correlation spectrum. If various protons in a magnetic field resonate at different energy levels the observation of spin exchange between these protons is permitted [35].

During the past two decades, a wide range of materials has been analyzed by this or similar 2D techniques [64–69]. A typical example of a semi-quantitative interpretation of 2D spin-exchange CRAMPS experiment can be demonstrated on the investigation of the extent of mixing of multicomponent polymer blends based on semicrystalline polycarbonate (*PC*) and semicrystalline *PEO* (*PC-PEO*) [48]. As displayed in Fig. 5, cross-peaks indicating proximity between aromatic and methyl protons of *PC* molecules are fully equilibrated after 250 μs. This indicates the shortest interatomic distance between methyl and aromatic protons within one monomer unit of *ca.* 0.3 nm, which is in accord with structural models of *PC*. The first cross-signals indicating dipolar interaction between methylene protons of *PEO* and both-type protons of the *PC* molecule are perceptible in a 2D spectrum measured with 500 μs mixing time. This indicates an intimate mixing of the amorphous phases of both components. However, even after 10 ms mixing time, the signals in the 2D spectrum are not yet completely equilibrated, which means that heterogeneities larger than several nanometers are present in this system. This heterogeneities correspond to crystallites of *PEO* and *PC*. These results reveal the complex morphology and arrangement of both polymer components.

Domain Size in the Diblock Copolymer PE-PEO

Detailed analysis of the spin-diffusion built-up curve demonstrates the power of this technique. The diblock copolymer *PE-PEO* provides an example of a system with relatively complicated morphology, because both polymer blocks are semicrystalline. Both chemically distinct $-CH_2-$ units are resolved in a 2D CRAMPS spectrum (see Fig. 6) allowing to obtain accurate spin-exchange data (see Fig. 7). From the known composition of the copolymer and taking into account the fraction of crystalline and amorphous phase, we estimated the effective spin diffusion coefficient $D_{eff} = 0.4\,nm^2\,ms^{-1}$. In the first approximation, a simple two-component phase-separated system consisting of domains of *PE* and *PEO* was assumed. From the analysis of spin diffusion by this model (dashed line in Fig. 7) it was found that both components form domains with the average size *ca.* 6.5 nm in lamellar morphology ($\varepsilon = 1$). Consequently the long period is 13 nm. From the known number-average polymerization degree (X_n) 170 and 109 of *PE* and *PEO*

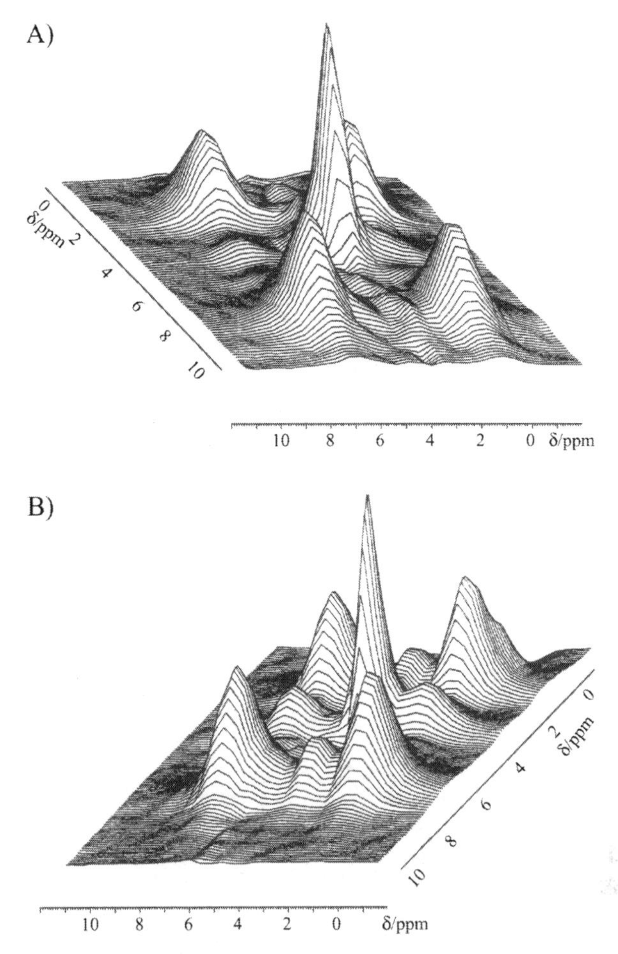

Fig. 5. 2D ^{1}H CRAMPS spectrum of the polymer blend *PC-PEO* measured with a 500 μs (A) and 10 ms mixing period (B)

chains it is clear that lamellae do not consist of extended chains, the theoretical lengths of which are 41 and 30 nm. This indicates extensive folding of both types of polymer chains. However, the model of a simple two-phase system does not quite fit the experimental dependence. That is why we modified the numerical simulation of the spin diffusion so that the single-spin-diffusion process was extended to a double-spin-diffusion process. The slow and fast spin-diffusion processes are superimposed and take place simultaneously. We propose that the first process apparent at the beginning of spin diffusion is a fast magnetization transfer between both components intimately mixed in the amorphous phase. For well-mixed polymer chains, cylindrical morphology and thus a dimensionality of $\varepsilon = 2$ is the best choice. The slow spin-diffusion process corresponds to the magnetization transfer involving substantially larger crystallites of both polymer chains. Simulation of spin exchange process according to this model revealed that

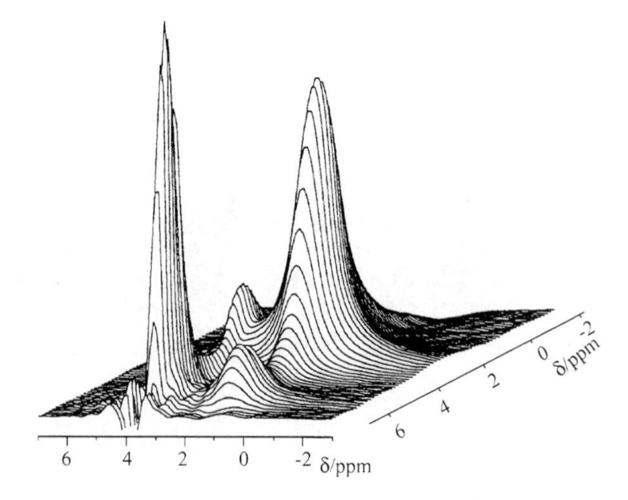

Fig. 6. 2D ^{1}H CRAMPS spectrum of the diblock copolymer *PE-PEO* measured with a 10 ms mixing period

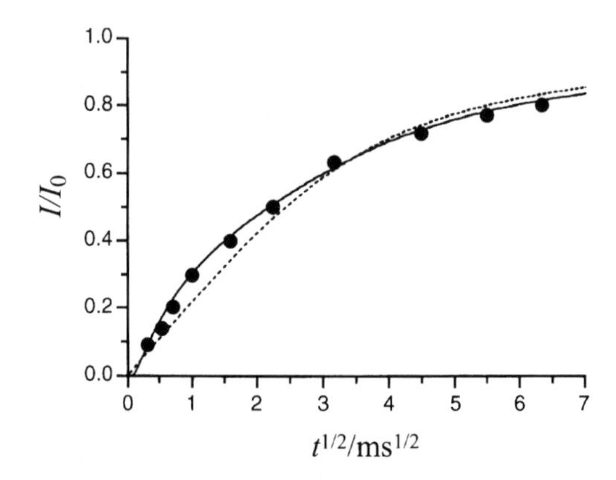

Fig. 7. Experimental (dots) and simulated (dashed and solid lines) spin-diffusion dependences of cross-peak intensity on mixing time obtained for the diblock copolymer *PE-PEO*

both components (*PE* and *PEO*) form relatively small domains with diameters *ca.* 1.0 and 0.5 nm in the amorphous phase, while crystallites are substantially larger, *ca.* 6.0 nm. The long period is *ca.* 13.5 nm, which exactly corresponds with the long period calculated for the simple model discussed above. This self-consistence of the obtained results proves the reliability of the applied model.

Clustering of Surface Hydroxyls in Siloxane Networks

Characterization of structure and geometry of organic polymer systems is not the only use of spin-exchange experiments. Recently, the simulation of a build-up curve has been used to investigate membrane peptide topology [70], as well as

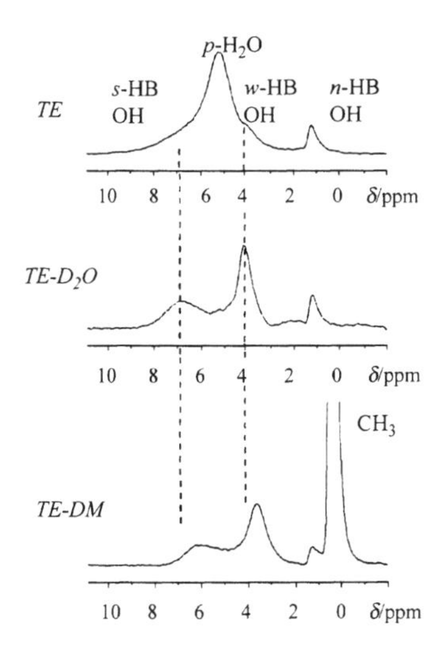

Fig. 8. ^{1}H CRAMPS NMR spectra of silica networks *TE*, *TE*-D_2O, and *TE-DM*

hydrogen-bonding and the size of clusters of various hydroxy groups in organically-modified polysiloxane networks [71].

Polysiloxane and polysilsesquioxane networks are short-range-ordered materials intermediate between the completely crystalline cristobalite and the least ordered silicate glasses [72–75]. Four basic types of hydroxyl groups have been detected in the CRAMPS spectrum even in hydrated silica gel (*TE*) prepared by polycondensation of tetraethoxysilane (see Fig. 8): strongly hydrogen-bonded (*s*-HB OH) and weakly hydrogen-bonded hydroxyl groups (*w*-HB OH), physisorbed water (*p*-H_2O) and non-hydrogen-bonded silanols (*n*-HB OH) at 7.0, 4.3, 5.2, and 1.4 ppm. Well-resolved signals in CRAMPS spectra of partially deuterated silica network (*TE*-D_2O) excludes fast chemical exchange between various hydroxyl sites and reflect a limited number of possible arrangements of strongly and weakly hydrogen-bonded OH. As the position of these ^{1}H NMR signals reflects the strength of hydrogen bonds it is clear that the network of hydrogen bonds is neither uniform nor random. Rather, both types of hydroxyl groups form several structures differing by the most probable length of hydrogen-bonds.

The presence of methyl substituents in the modified network prepared by co-condensation of tetraethoxysilane and dimethyldiethoxysilane (*TE-DM*) moves chemical shifts of signals of both strongly and weakly hydrogen-bonded hydroxyls toward higher field (see Fig. 8). Although this indicates a decrease in the hydrogen-bond strength, the replacement of silanol sites by methyls does not completely destroy the formation of hydrogen-bonding networks involving silanols and water molecules.

We have studied ^{1}H–^{1}H interatomic distances and the size of various OH clusters in detail [71]. Even though it is generally accepted that silanols reflected by the signal at 1.4 ppm are isolated and "water-inaccessible", the appearance of a cross-peak correlating it to the 5.2 ppm signal clearly proves spin exchange between these silanols and physisorbed water (see Fig. 9A). Their mutual distance is smaller

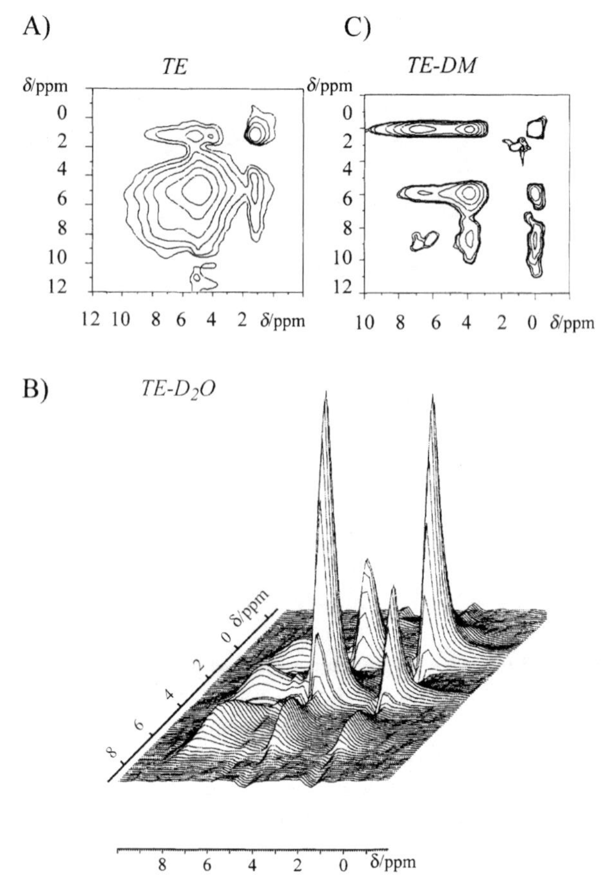

Fig. 9. 2D ^{1}H spin-exchange CRAMPS spectra of *TE*, *TE*-D_2O, and *TE-DM* systems measured at 20 ms spin-diffusion mixing time (A, B, and C)

than 0.4 nm. More detailed structure information was obtained for a partly deuterated sample (*TE*-D_2O). Quantitative data were derived from the simulation of spin-diffusion process for a two-component system with an interface with variable diffusivity and dimensionality ($\varepsilon = 1, 2, 3$). An effective spin diffusion coefficient was estimated according to *Assink*'s relation [76] from the ^{1}H T_2 relaxation constant. Basic parameters used for spin diffusion simulation are listed in Table 1.

The clusters of strongly and weakly hydrogen-bonded OH in the system *TE*-D_2O form relatively large regions and non-hydrogen-bonded silanols are dipolar-coupled with both types of these protons (see Fig. 9B). The absence of the interface following from the analysis of the spin-diffusion dependences (see Fig. 10A) indicates that all three types of OH are in mutual contact and the majority of strongly and weakly hydrogen bonded clusters are located at the surface. As the cross-peak intensity correlating strongly and weakly hydrogen-bonded OH does not tend to reach the theoretical equilibrium intensity, we suggest that a part of weakly hydrogen-bonded OH protons is quite isolated. From the best fits, employing our

Table 1. Transverse relaxation time, T_2, and calculated spin-diffusion coefficients, D

System	Hydroxyl type	T_2 ms	D^a $mm^2\,ms^{-1}$
TE	*p*-H_2O	0.55	0.05
	s-HB OH	0.45	0.06
	w-HB OH	1.70	0.02
	n-HB OH	20.0	0.001
TE-DM	*s*-HB OH	0.60	0.05
	w-HB OH	1.80	0.02
	$-CH_3$	0.25	0.11

[a] Calculated by *Assink*'s method [76] $D_{eff} = 2(r_0)^2/T_2$, where r_0 is the *van der Waals* radius

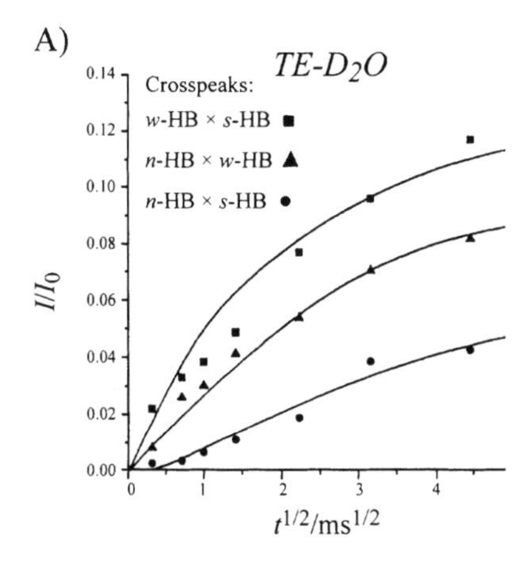

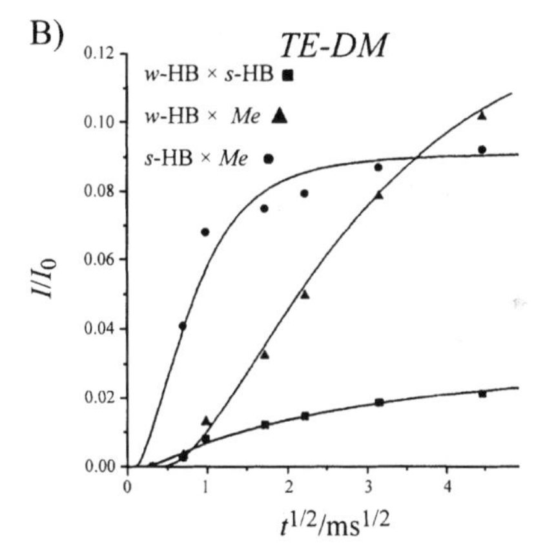

Fig. 10. Experimental (dots) and simulated (line) spin-diffusion curves: dependences of cross-peak intensity on mixing time

assumption about main dimensionality ($\varepsilon = 2$) we estimated the average size of hydroxyl clusters. Thus, it seems to be reasonable to conclude that weakly hydrogen-bonded OH groups form the largest clusters with a maximum diameter of about 1.5–2.0 nm, while clusters of strongly-hydrogen bonded OH are smaller with diameters below 1.0 nm. The smallest size was obtained for non-hydrogen-bonded silanols (*ca.* 0.5–0.4 nm) indicating that they can be formed by two geminal silanols and/or by two neighboring single silanols, which are in a geometry inappropriate to form a hydrogen bond.

The size of domains in a modified network *TE-DM* of strongly and weakly hydrogen-bonded OH groups is approximately the same as in the case of a net silica network. Methyl groups are in close contact with both types of clusters confirming that the presence of a small number of methyl units at the surface does not interrupt the hydrogen-bonding network. A high rate of equilibration of cross-peak intensity correlating methyls and strongly hydrogen-bonded OH (see Fig. 10B) reflects their intimate mixing. The calculated size of the dimethylsiloxane domains of about 1 nm indicates that the dimethylsiloxane monomer units occur in pairs. Significant interface was found between methyls and weakly hydrogen-bonded silanols. We propose that this interface reflects a portion of methyl groups which are surrounded only by strongly hydrogen-bonded OH.

Determination of ^1H–^1H Interatomic Distance

Significant progress has recently been made in the improvement in homonuclear dipolar decoupling sequences [27, 28, 77, 78]. Resolutions sufficient to distinguish signals having chemical shift differences as small as 0.5 ppm have been achieved. As ^{1}H NMR signals are usually detected indirectly, three-dimensional (3D) techniques [29, 30] have been used to measure ^{1}H–^{1}H correlations, although quite recent applications of windowed PMLG scheme has reduced the dimensionality of these experiments [31]. For instance, a 3D ^{1}H–^{1}H–^{13}C correlation experiment [30] correlated two high-resolution ^{1}H spectra with the ^{13}C NMR spectrum. Thereby the nearest ^{1}H–^{1}H interatomic distances can be selectively probed for each carbon resolved in the ^{13}C NMR spectrum. From this follows the possibility to measure ^{1}H–^{1}H correlations in solids providing structural constrains similar to those used to determine structures in liquid-state NMR.

On the basis of these facts, recently we have performed a detailed analysis of a spin-exchange process leading to the determination of the nearest ^{1}H–^{1}H interatomic distances [79]. We have found that difference in local molecular motions of various groups, even in virtually motionally homogeneous highly rigid systems (crystals of small organic molecules) substantially affect simulation of spin-exchange process. Our analysis of the spin-exchange process has been carried out on crystalline α-glycine, which is a highly suitable sample. The resulting ^{1}H CRAMPS spectrum is well resolved with a line-width of *ca.* 0.5–0.8 ppm and exhibits splittings of signal of α-protons reflecting their magnetic nonequivalence (different electrostatic potential charges [80]; see Fig. 11). Since fast rotation of NH_3^+ groups is assumed from the shape of the ^{1}H CRAMPS NMR signal reflecting a partial averaging of ^{14}N–^{1}H dipolar interactions [81, 82], NH_3^+ protons are considered to be a relatively mobile ^{1}H moieties, while a pair of α-protons is more rigid.

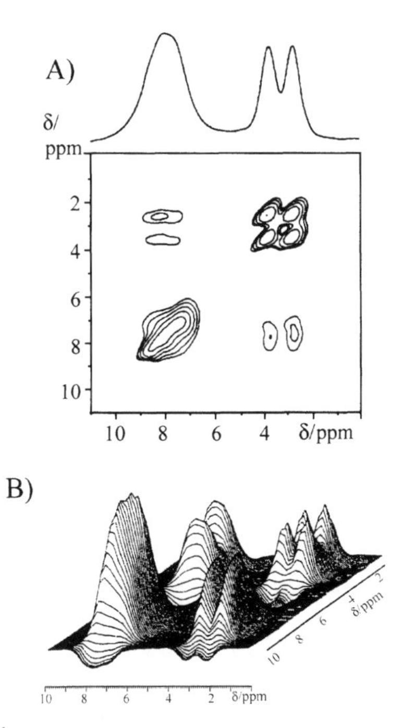

Fig. 11. 2D spin-exchange ^{1}H CRAMPS spectra measured at various mixing times, (A, 50 μs and B, 300 μs)

In practice it is impossible to detect this difference in molecular motion from a simple static ^{1}H NMR spectrum or a standard inversion recovery T_1 relaxation experiment, because fast ^{1}H spin flip-flop leads to a homogeneously broadened ^{1}H NMR spectrum (line-width *ca.* 50 kHz) and single-component relaxation behavior.

Evolution of cross-peak intensities as a function of $\sqrt{t_m}$ is displayed in Fig. 12. Analysis of the linear part of spin-exchange built-up curves was performed according to Eq. (3), which has been recently used [83] to calibrate spin-diffusion coefficients by analysis of intramonomer polarization transfers involving CH_3 and CH_2 protons in a polyisobutylene sample. It has been shown that the main pathway for magnetization exchange occurs only within the monomer unit of a single chain [83]. That is why we assume that intermolecular spin-exchange involving the shortest interatomic ^{1}H–^{1}H distance within one molecule (see Fig. 13) is the dominant process during the initial step of polarization transfer.

Dependences of cross-peak intensities on mixing time as depicted in Fig. 12A clearly reflect a substantial difference in the spin-exchange rate between two nonequivalent α-protons and between α-protons and NH_3^+ groups. There is a remarkable decrease in intensity of the signal correlating nonequivalent α-protons during later stages of polarization transfer, which clearly indicates a substantially smaller spin-exchange rate constant driving polarization transfer between αH and NH_3^+ protons. To determine the spin-exchange coefficient exactly and to avoid the decrease of cross-peak signal intensity, polarization transfer between nonequivalent α-protons was analyzed separately (see Fig. 12B). On the basis of the known

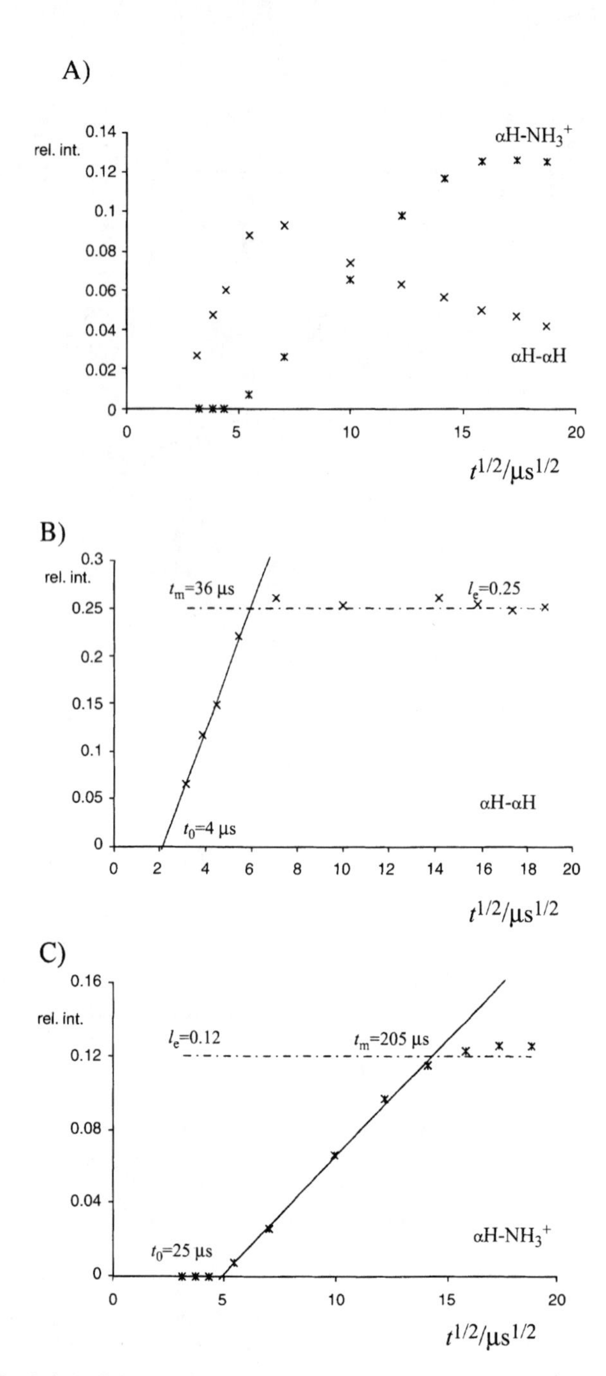

Fig. 12. (A) Evolution of the correlation signal intensity as a function of $\sqrt{t_m}$; (B) Experimental spin-exchange built-up curve and simulation of its linear part for polarization transfer between nonequivalent α-protons; (C) Experimental spin-exchange built-up curves and simulation of the linear parts for polarization transfer between nonequivalent α-H and NH_3^+ protons in glycine

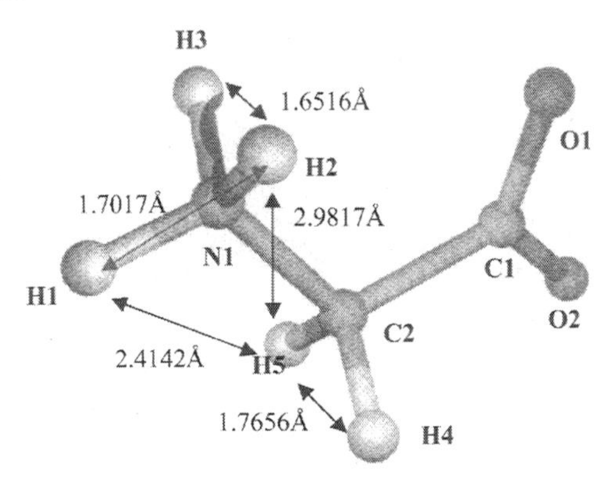

Fig. 13. Geometry of glycine molecules in the crystal unit cell obtained from neutron diffraction data [84]

interatomic distance between α-protons (1.77 nm) [84], and mixing time ($t_m = 32\ \mu s$), which is necessary to achieve the equilibrium intensity of the corresponding cross-peak, the spin-exchange constant $D = 0.77\ nm^2\ ms^{-1}$ was derived. The calculated spin-exchange coefficient corresponds to the values generally reported in literature for rigid organic solids, $D = 0.7–0.8\ nm^2\ ms^{-1}$.

In the next step, we analyzed the spin-exchange process between αH and NH_3^+ protons. Generally it is accepted that in a system with uniform internal mobility, the determined spin-exchange constant is the same for all spin-pairs. Thus, assuming that the motion of the system is homogeneous (*i.e.* applying $D = 0.77\ nm^2\ ms^{-1}$), the analysis of spin-exchange processes between α-H and NH_3^+ protons should reveal interatomic distances corresponding to *ca.* 0.24–0.25 nm (neutron diffraction data [84]).

The long delay at the beginning of the polarization transfer process during which no off-diagonal cross-peak correlating αH and NH_3^+ protons evolves (see Fig. 12C) probably corresponds to back polarization exchange within each ^1H moiety. A similar, but not quite the same phenomenon has been observed by *de Groot et al.* [85] in heteronuclear polarization transfer experiment probing $^1H–^{13}C$ interatomic distances in tyrosine. Due to this fact we believe that this delay-time can be substracted from the experimentally determined equilibrium mixing times ($t_m = 205\ \mu s$). However, even with this correction and using equilibrium mixing times $t_m = 181\ \mu s$ the calculated distance between NH_3^+ and αH is very large ($r = 0.43$ nm). This deviation cannot be simply explained by relayed coherence transfer, because assuming the largest coherence pathway (αH-1 $\rightarrow$ αH-2 $\rightarrow$ NH-1 $\rightarrow$ NH-2) the time which would be necessary to achieve equilibrium intensity is only $t_m = 32 + 58 + 29 = 119\ \mu s$. From this, it is clear that the rotation of the NH_3^+ group predominantly affects the rate of spin-exchange. The local molecular motion averages dipolar interactions and consequently the corresponding spin-exchange rate constant is reduced. The corresponding spin-exchange rate constant driving polarization transfer between NH_3^+ and αH is scaled down by a factor of approximately 3.5 ($D^* = 0.24\ nm^2\ ms^{-1}$).

Although the difference in internal motions cannot be observed by standard ^1H NMR experiments, analysis of ^2H NMR line-shapes of selectively deuterium

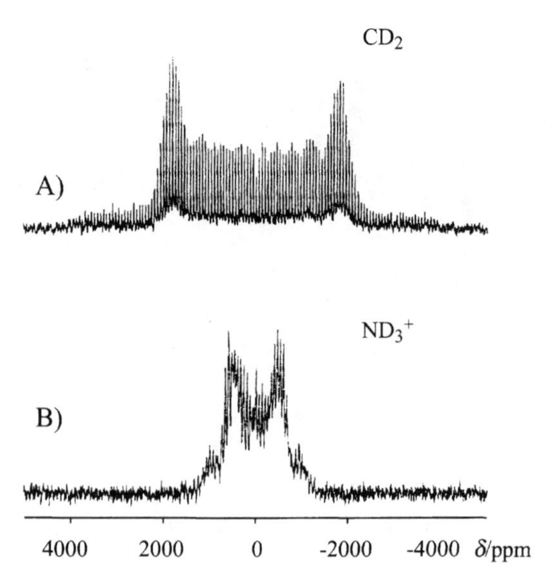

Fig. 14. ^{2}H MAS spectra of selectively deuterated samples: (A) $NH_3^+CD_2COO^-$, (B) $ND_3^+CH_2COO^-$

labeled samples (see Fig. 14) directly proved a difference in the local mobilities of CD_2 and ND_3^+ moieties. The ^{2}H quadrupole splitting QCC is approximately three times smaller for the amino group compared with the methylene one, which nicely correlates to the observed decrease of the spin-exchange constant: $D^*/D \cong QCC(ND_3^+)/QCC(CD_2) \cong 1/3.5$.

Hence, the relative spin-exchange rate constants can be evaluated on the basis of the knowledge of the ^{2}H line-shape of specific sites.

Determination of ^{1}H–^{1}H Interatomic Distance Through ^{13}C–^{13}C Correlation

The simple ^{1}H–^{1}H correlation experiment just discussed is sufficient only for small molecular systems. For larger macromolecules, spectral resolution has to be increased, *e.g.* by application of 3D ^{1}H–^{1}H–^{13}C techniques [30] as mentioned above or 2D ^{13}C–^{13}C correlation mediated by ^{1}H–^{1}H spin-exchange [55, 86] (see Fig. 15). Although the later experiment looks very promising as it should allow to determine short ^{1}H–^{1}H interatomic distances (see Fig. 16A), it has been used so far only to assign ^{15}N resonance in uniformly labeled biological solids [87], to study the degree of mixing of principal and secondary phase of $Zn(O_3PC_2H_4COOH)0 \cdot 5C_6H_5NH_2$ through ^{31}P–^{31}P correlation [88], or to evaluate the degree of phase separation and domain size in selectively labeled polymer blends [86]. This means that predominantly distances in the nonometer length scale have been probed. The impossibility to determine shorter distances by this experiment follows from the application of a cross-polarization (CP) step for coherence transfer. It is known that ^{1}H–^{1}H spin-exchange occurs not only for z magnetization but also for spin-locked transverse ^{1}H magnetization, for instance under *Hartman-Hahn* CP [89, 90]. Under sample rotation, the locking field scales the dipolar coupling by a factor 1/2, which causes

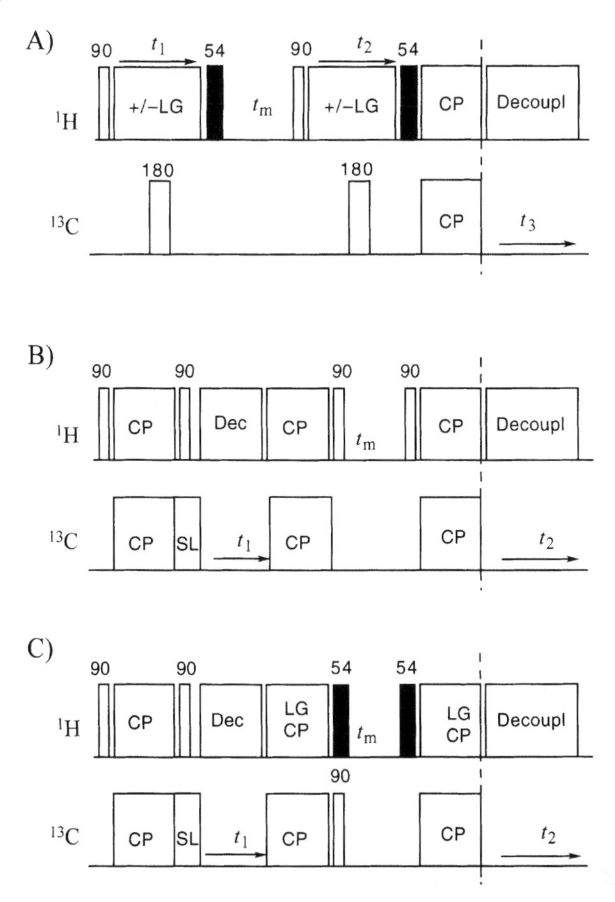

Fig. 15. (A) Pulse scheme of 3D spin-exchange experiment with *Lee-Goldburg* decoupling during both indirect ^{1}H detection period [30]; (B) 2D ^{13}C–^{13}C correlation experiment mediated by ^{1}H–^{1}H spin-exchange [54]; (C) Modification of B by *Lee-Goldburg*-CP and magic angle pulses (SL – spin lock, CP – cross-polarization, LG – *Lee-Goldburg*, Dec. – heteronuclear decoupling)

slowing down of the spin-exchange process by one half during CP as compared with the standard spin-exchange rate taking part in z direction. However, the spin exchange is not completely quenched. As the spin-exchange process in rigid solids is very fast (several tens of microseconds is sufficient to achieve equilibrium), a high degree of equilibration is achieved even at very short cross-polarization times without any mixing time.

In the case of this 2D ^{13}C–^{13}C correlation experiment, the relevant CP steps complicating the evolution of cross-peak intensity are the second and the third one. As the result of this undesired coherence transfer one can clearly see cross-peaks in the 2D spectrum of uniformly ^{13}C enriched alanine which was acquired with zero mixing time and very short cross-polarization periods (see Fig. 16B). Due to these artificial signals, the experimentally determined dependence of cross-peak intensity on mixing time is not suitable for accurate analysis (see Fig. 17A).

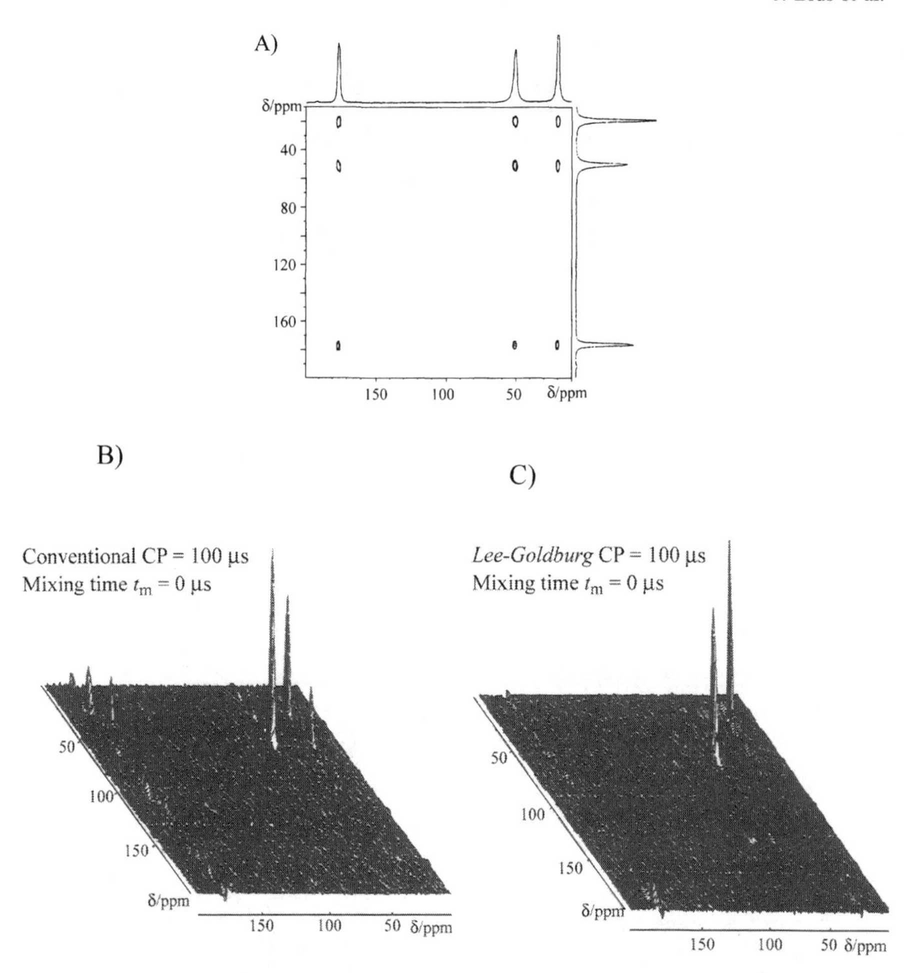

Fig. 16. 2D ^{13}C–^{13}C correlation spectra of U–^{13}C, ^{15}N Alanine measured at: (A) 100 ms cross-polarization (CP) time and 1 ms mixing time; (B) 100 ms CP time and zero mixing time; (C) 100 ms *Lee-Goldburg* CP time and zero mixing time

Spin-exchange during CP has to be suppressed and the artifacts have to be removed to be able to analyze the evolution of cross-peak intensity. A very promising tool to suppress ^{1}H–^{1}H dipolar interactions is provided by *Lee-Goldburg* irradiation [23]. Applying this technique, for instance during acquisition of the ^{13}C resonance, the ^{1}H–^{1}H spin-diffusion is suppressed and the resulting spectrum contains multiplets reflecting the number of *J*-coupled protons. It has been shown that the *Lee-Goldburg* irradiation can also be applied during the cross-polarization transfer step [85]. Observation of intense transient dipolar oscillation in the case of *Lee-Goldburg* CP confirms the suppression of ^{1}H–^{1}H spin-exchange. Recently, it has been shown that an analysis of this dipolar oscillation reflecting dipolar coupling between directly bonded ^{1}H–^{13}C spin pairs can be used to obtain accurate ^{1}H–^{13}C interatomic distances [85]. In the case of conventional on-resonance CP

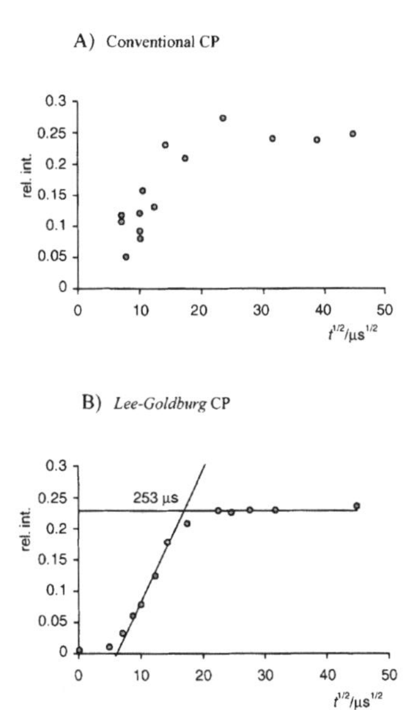

Fig. 17. Spin-exchange built-up curves obtained by application of conventional experiment (A) and by modified pulse sequence with *Lee-Goldburg*-CP (B)

such dipolar oscillation is almost completely destroyed as a result of fast ^{1}H–^{1}H spin exchange involving many ^{1}H spins. By application of *Lee-Goldburg* CP for the second and the third polarization transfer in the 2D ^{13}C–^{13}C correlation pulse sequence the unwanted spin-exchange was suppressed. The original 90° ^{1}H pulses (exactly, the third and the fourth one) have been replaced by magic angle pulses.

In the resulting 2D spectrum measured with *Lee-Goldburg* CP and without any mixing period the unwanted artificial cross-peaks are completely canceled. In addition, signal intensities of diagonal signals are not distorted when applying *LG*-CP. From the obtained dependence of cross-peak intensity on mixing time (see Fig. 17B), which is now suitable for accurate analysis, one can determine the desired information about ^{1}H–^{1}H interatomic distance with high accuracy according to the procedures mentioned in the previous sections.

Conclusion

In this contribution, we briefly summarized basic experimental techniques leading to averaging of ^{1}H–^{1}H dipolar interaction and allowing thus to obtain highly resolved ^{1}H NMR spectra of organic solids. Spin-diffusion experiments providing an interesting probe to geometry and structure of several systems were introduced and analysis of resulting data was discussed. The power of theses 2D spin-exchange experiments was demonstrated on several complex systems. At first, the degree of mixing of polymer components was evaluated for the semicrystalline blend *PC-PEO* and a further detailed analysis of complex spin-diffusion process led to an estimate of the size of polymer domains in the diblock copolymer *PE-PEO*. The same approach was used to investigate clustering of surface hydroxyls on a silica network and to determine the average size of various hydroxyl clusters. It was also shown that the same experiments and procedures can be used to determine very short interatomic ^{1}H–^{1}H distances, although the variation in internal motions has to be carefully investigated and described. Finally we discussed the possibility of the application of 2D ^{13}C–^{13}C correlation experiments exploiting ^{1}H–^{1}H spin exchange to determine structural constrains.

Experimental

NMR Spectroscopy

NMR spectra were measured using a Bruker DSX 200 NMR spectrometer (Karlsruhe, Germany) in 4 and 7 mm ZrO_2 rotors at frequencies 50.33 and 200.14 MHz (^{13}C and ^{1}H, respectively). For acquisition of ^{1}H MAS NMR spectra, the spinning frequency was 0–16 kHz and the strength of the B_1 field 62.5 kHz ($\pi/2$ pulse 4 µs). 1D CRAMPS spectra at slow MAS (2 kHz) were acquired using the BR-24 pulse sequence [18]. A 2D spin-exchange experiment proposed by *Caravatti et al.* [63] was used to observe ^{1}H–^{1}H correlation. In direct and indirect detection periods the BR24 pulse sequence was used. Spin-diffusion mixing times varied from 0.1 to 40 ms. The intensity of the B_1 field was 140 kHz ($\pi/2$ pulse 1.8 µs) and small and large windows were 1.0 and 3.8 µs, respectively. The ^{1}H scale was calibrated with glycine as an external standard (low-field NH_3^+ signal at 8.0 ppm and the high field α-H signal at 2.8 ppm). 2D ^{13}C–^{13}C correlation spectra were obtained with a pulse sequence proposed by *de Groot et al.* and *Spiess et al.* [55, 86] at 11 kHz. The strength of the B_1 field applied for the *Lee-Goldburg* cross-polarization was 83 kHz, with ^{1}H resonance offset 64.6 kHz.

Materials

The *PC-PEO* blend was prepared from commercial-grade Bisphenol A polycarbonate SINVET 251 (ENI, Italy) with a weight-average molecular weight (M_w) of 24000 and a number-average molecular weight (M_n) of 9600, and from *PEO* ($M_w = 6 \times 10^5$) produced by BDH Chemicals, Ltd. (UK). The sample of *PC-PEO* blend was obtained by dropwise precipitation from a $CHCl_3$ solution (2% w/w) into pentane, slow evaporation of the solvents at room temperature and subsequent heating in a vacuum oven at 85°C for 1 h.

The block-copolymers *PE-PEO* $M_n(PE) = 4700$, $M_n(PEO) = 4700$ was used without any purification as purchased from Polymer Source, Inc.

Siloxane materials *TE* and *TE-DM* were prepared by acid-catalyzed sol-gel polycondensation of mixtures: tetraehoxysilane (*TEOS*)/C_2H_5OH/H_2O/HCl and tetraethoxysilane/dimethyldiethoxysilane (*DMDEOS*)/C_2H_5OH/H_2O/HCl in mole ratios 1/4.50/3/0.03 and 0.75/0.25/4.50/3/0.03, respectively. *TEOS* and *DMDEOS* were purchased from Wacker-Chemie GmbH., Germany. HCl was added to a mixture of alkoxysilanes with ethanol. The resulting mixture (*ca.* 10 g) was stirred for 30 min and subsequently poured onto a *Petri* dish (5.5 cm in diameter). Polycondensation then took place under laboratory conditions. After a year, the products were finely powdered and placed into an air-conditioned box (relative humidity – RH = 55%, $t = 25$°C) for one month. Partially deuterated samples were obtained by simple exchange with deuterium oxide at laboratory temperature and pressure in a close vessel containing a dish with D_2O. After the deuteration procedure the samples were not subsequently dried. Deuterium exchange periods were 24 h.

Glycine and U–^{15}N,^{13}C Alanine were used as purchased from Aldrich.

Acknowledgement

The authors thank the Grant Agency of the Academy of Sciences of the Czech Republic (grant IAB4050203, IAA4050208 and grant AVOZ4050913) for financial support.

References

[1] Mehring M (1983) In: Principles of high resolution NMR in solids, Springer, Berlin
[2] Schmidt-Rohr K, Spiess HW (1994) In: Multidimensional solid-state NMR and polymers, Academic Press, London
[3] Lowe IJ (1959) Phys Rev Lett **2**: 285
[4] Marciq MM, Waugh JS (1979) J Chem Phys **70**: 3300
[5] Samoson A, Tuherm T (1999) In: Proceedings, the 1-st alpine conference on solid-state NMR, Chamonix-Mont Blanc, France
[6] Bjorholm T, Jakobsen HJ (1989) J Magn Reson **84**: 204
[7] Bielecki A, Burum DP (1995) J Magn Reson A **116**: 215
[8] Aguilar-Parrila F, Wehrle B, Braunling H, Limbach HH (1990) J Magn Reson **87**: 592
[9] Brus J (2000) Solid State Nucl Magn Reson **16**: 151
[10] Waugh JS, Huber LM, Haeberlen U (1968) Phys Rev Lett **20**: 453
[11] Haeberlen U, Waugh JS (1968) Phys Rev Lett **20**: 180
[12] Farrar TC (1990) Concepts Magn Reson **2**: 55
[13] Smith SA, Palke WE, Gerig JT (1993) Concepts Magn Reson **5**: 151
[14] Gerstein BC, Pembleton RG, Wilson RC, Ryan LM (1977) J Magn Reson **66**: 361
[15] Maciel GE, Bronnimann CE, Hawkins B (1990) Adv Magn Reson **14**: 125
[16] Dec SF, Bronnimann CE, Wind RA, Maciel GE (1989) J Magn Reson **82**: 454
[17] Rhim WK, Elleman DD, Vaughan RW (1973) J Chem Phys **59**: 3740
[18] Burum DP, Rhim WK (1979) J Chem Phys **71**: 944
[19] Bronnimann CE, Hawkins BL, Zhang M, Maciel GE (1988) Anal Chem **60**: 1743
[20] Hafner S, Spiess HW (1996) J Magn Reson A **121**: 160
[21] Demco DE, Hafner S, Spiess HW (1995) J Magn Reson **116**: 36
[22] Hafner S, Spiess HW (1997) Solid State Nucl Magn Reson **8**: 17
[23] Lee M, Goldburg WI (1965) Phys Rev A **140**: 1261
[24] Bielecki A, Kolbert AC, Levitt MH (1989) Chem Phys Lett **155**: 341
[25] Bielecki A, Kolbert AC, de Groot HJM, Griffin RG, Levitt MH (1990) Adv Magn Reson **14**: 111
[26] Lesage A, Steuernagel S, Emsley L (1998) J Am Chem Soc **120**: 7095
[27] Lesage A, Duma L, Sakellariou D, Emsley L (2001) J Am Chem Soc **123**: 5747
[28] Vinaogradov E, Madhu PK, Vega S (1999) Chem Phys Lett **314**: 443
[29] Ramamoorthy A, Gierasch LM, Opella SJ (1996) J Mag Reson Ser B **111**: 81
[30] Sakellariou D, Lesage A, Emsley L (2001) J Am Chem Soc **123**: 5604
[31] Vinogradov E, Madhu PK, Vega S (2002) Chem Phys Lett **354**: 193
[32] Levitt MH, Kolbert AC, Bielecki A, Ruben DJ (1993) Solid State NMR **2**: 151
[33] Bloembergen N (1949) Physica **15**: 386
[34] Zhang S, Meier BH, Ernst RR (1992) Phys Rev Lett **69**: 2149
[35] Abragam A (1961) In: The Principles of Nuclear Magnetism, Oxford University Press, London
[36] Clauss J, Schmidt-Rohr K, Spiess HW (1993) Acta Polym **44**: 1
[37] Cheung TTP (1999) J Phys Chem B **103**: 9423
[38] Demco DE, Johanson A, Tegenfeldt J (1995) Solid State Nucl Magn Reson **4**: 13
[39] VanderHart DL, McFadden GB (1996) Solid State Nucl Magn Reson **7**: 45
[40] Campbell GC, Vander Hart DL (1992) J Magn Reson **96**: 69
[41] Mellinger F, Wilhelm M, Spiess HW (1999) Macromolecules **32**: 4686
[42] Holstein P, Monti GA, Harris RK (1999) Phys Chem Chem Phys **1**: 3549
[43] Cheung TTP (1981) Phys Rev B **23**: 1404
[44] Cheung TTP, Gerstein BC (1981) J Appl Phys **52**: 5517
[45] Chin YH, Kaplan S (1994) Magn Res in Chem **32**: S53
[46] Weigand F, Demco DE, Blümich B, Spiess HW (1996) J Magn Reson Ser A **120**: 190
[47] Hu WG, Schmidt-Rohr K (2000) Polymer **41**: 2979
[48] Brus J, Dybal J, Schmidt P, Kratochvíl P, Baldrian J (2000) Macromolecules **33**: 6448

[49] Spiegel S, Schmidt-Rohr K, Böffel C, Spiess HW (1993) Polymer **34**: 4566
[50] Lehmann SA, Meltzer AD, Spiess HW (1998) J Polym Sci, Part B, Polym Phys **36**: 693
[51] Clauss J, Schmidt-Rohr K, Adam A, Boeffel C, Spiess HW (1992) Macromolecules **25**: 5208
[52] Cheung MK, Wang J, Zheng S, Mi Y (2000) Polymer **41**: 1469
[53] Brus J, Dybal J, Sysel P, Hobzová R (2002) Macromolecules **35**: 1253
[54] Clayden NJ, Nijs CL, Eeckhaut GJ (1997) Polymer **38**: 1011
[55] Wilhelm M, Feng H, Tracht U, Spiess HW (1998) J Magn Reson **134**: 255
[56] Mirau PA, Shu Yang (2002) Chem Mater **14**: 249
[57] Goldman M, Shen L (1996) Phys Rev **144**: 321
[58] VanderHart DL, Feng Y, Han CC, Weiss RA (2000) Macromolecules **33**: 2206
[59] Beshah K, Molnar LK (2000) Macromolecules **33**: 1036
[60] Schmidt-Rohr K, Clauss J, Blumich B, Spiess HW (1990) Magn Res in Chem **28**: 3
[61] Caravatti P, Lewitt MH, Ernst RR (1986) J Magn Reson **68**: 323
[62] Caravatti P, Neuenschwander P, Ernst RR (1986) Macromolecules **19**: 1889
[63] Caravatti P, Neuenschwander P, Ernst RR (1985) Macromolecules **18**: 119
[64] Nouwen J, Adriaensens P, Gelan J, Verreyt G, Yang Z, Geise HJ (1994) Macromol Chem Phys **195**: 2469
[65] Belfiore LA, Graham H, Veda E, Wang Y (1992) Polym Int **28**: 81
[66] Schaller T, Sebald A (1995) Solid State NMR **5**: 89
[67] Babonneau F, Gualandris V, Maquet J, Massiot D, Janicke MT, Chmelka BF (2000) J Sol-Gel Sci & Techn **19**: 113
[68] Mirau PA, Heffner SA, Schilling M (1999) Chem Phys Lett **313**: 139
[69] Heffner SA, Mirau PA (1994) Macromolecules **27**: 7283
[70] Huster D, Yao X, Hong M (2002) J Am Chem Soc **124**: 874
[71] Brus J, Dybal J (2002) Macromolecules (in press)
[72] Chuang IS, Maciel GE (1997) J Phys Chem B **101**: 3052
[73] Kinney DR, Chuang IS, Maciel GE (1993) J Am Chem Soc **115**: 6786
[74] Changhua CL, Maciel GE (1996) J Am Chem Soc **118**: 5103
[75] Chuang IS, Kinney DR, Maciel GE (1996) J Am Chem Soc **118**: 8695
[76] Assink RA (1978) Macromolecules **11**: 1233
[77] Lesage A, Emsley L (2001) J Magn Reson **148**: 449
[78] Sakellariou D, Lesage A, Hodgkinson P, Emsley L (2000) Chem Phys Lett **319**: 253
[79] Brus J, Petrickova H, Dybal J (2002) Solid State NMR (submitted)
[80] Kimura H, Nakamura K, Eguchi A, Sugisawa H, Deguchi K, Ebisawa K, Suzuki E, Shoji A (1998) J Mol Struct **447**: 247
[81] Naito A, Root A, McDowell CA (1991) J Phys Chem **95**: 3578
[82] Jackson P, Harris RK (1995) J Chem Soc Faraday Trans **91**: 805
[83] Wang X, White JL (2002) Macromolecules **35**: 3795
[84] Power LF, Turner KE, Moore FH (1976) Acta Crystallogr **B32**: 11
[85] van Rossum BJ, de Groot CP, Ladizhansky V, Vega S, de Groot HJM (2000) J Am Chem Soc **122**: 3465
[86] Mulder FM, Heinen W, van Duin M, Lugtenburg J, de Groot HJM (1998) J Am Chem Soc **120**: 12891
[87] Wei Yufeng, Ramamoorthy A (2001) Chem Phys Lett **342**: 312
[88] Massiot D, Alonso B, Fayon F, Fredoueil F, Bujoli B (2001) Solid State Sci **3**: 11
[89] Hartmann SR, Hahn EL (1962) Phys Rev **128**: 2042
[90] Pines A, Gibby MG, Waugh JS (1973) J Chem Phys **59**: 569

SpringerChemistry

Microchimica Acta

An International Journal on Micro and Trace Analysis